FOOD AND BEVERAGE CONSUMPTION AND HEALTH

FOOD

THE VITAL INGREDIENT

FOOD AND BEVERAGE CONSUMPTION AND HEALTH

Additional books and e-books in this series can be found on Nova's website under the Series tab.

FOOD AND BEVERAGE CONSUMPTION AND HEALTH

FOOD

THE VITAL INGREDIENT

A. BRYAN HANLEY
AND
WENDY R. RUSSELL
EDITORS

nova
science publishers
New York

NOTICE TO THE READER

The Publisher has taken reasonable care in the preparation of this book, but makes no expressed or implied warranty of any kind and assumes no responsibility for any errors or omissions. No liability is assumed for incidental or consequential damages in connection with or arising out of information contained in this book. The Publisher shall not be liable for any special, consequential, or exemplary damages resulting, in whole or in part, from the readers' use of, or reliance upon, this material. Any parts of this book based on government reports are so indicated and copyright is claimed for those parts to the extent applicable to compilations of such works.

Independent verification should be sought for any data, advice or recommendations contained in this book. In addition, no responsibility is assumed by the publisher for any injury and/or damage to persons or property arising from any methods, products, instructions, ideas or otherwise contained in this publication.

This publication is designed to provide accurate and authoritative information with regard to the subject matter covered herein. It is sold with the clear understanding that the Publisher is not engaged in rendering legal or any other professional services. If legal or any other expert assistance is required, the services of a competent person should be sought. FROM A DECLARATION OF PARTICIPANTS JOINTLY ADOPTED BY A COMMITTEE OF THE AMERICAN BAR ASSOCIATION AND A COMMITTEE OF PUBLISHERS.

Additional color graphics may be available in the e-book version of this book.

Library of Congress Cataloging-in-Publication Data

ISBN: 978-1-53614-402-4

Published by Nova Science Publishers, Inc. † New York

CONTENTS

Introduction		**vii**
Acknowledgments		**xv**
Chapter 1	Global Food Issues and the Role of Technology *A. Bryan Hanley and Wendy R. Russell*	**1**
Chapter 2	Technologies Supporting Resilience in Primary Agriculture *Wendy R. Russell and A. Bryan Hanley*	**13**
Chapter 3	Physicochemical Characteristics of Food *Peter J. Wilde*	**37**
Chapter 4	Challenges to Food Safety and Quality: Technical Challenges *Mike Saltmarsh*	**53**
Chapter 5	Food Analysis and Food Standards: Raw Materials, Traceability and Authenticity *Martin Rose and Kathy Ridgway*	**69**
Chapter 6	Sustainability in a Global and Local Context *Liliya Serazetdinova and Jayne L. Brookman*	**91**

Chapter 7 Diet, Health and Chronic Disease **113**
Wendy R. Russell and A. Bryan Hanley

Chapter 8 Food Processing Manufacturing and Reformulation Processes **129**
Madalina Neacsu

Chapter 9 Consumer Awareness of Science and Technology Related to Food Retail and Mass Production **191**
A. Bryan Hanley

Chapter 10 Food Waste and By-Products Valorization **207**
Dimitris Charalampopoulos

Chapter 11 New Technologies: Industrial Biotechnology in Agri/Food **221**
Yvonne C. Armitage and A. Bryan Hanley

Chapter 12 The Chemistry and Physics of Eating **237**
Peter Lillford

Conclusion and Recommendations **247**

Editors Contact Information **257**

Index **259**

INTRODUCTION

PROBLEMS AND CHALLENGES

Food producers, processors, consumers, legislators and regulators all face considerable challenges over the coming years. For the first time in history we can see a horizon where there may not be enough food available to provide nutrition for all the people on the planet. In the past, there was inequality (which persists) but there was not a measureable gross shortage. To quote from Gandhi, “The world has enough for everyone's need, but not enough for everyone's greed.” That is no longer necessarily the case. Population increases and changes in eating habits and expectations, in part caused by the success of the food industry in providing new, attractive and safe food at a relatively low cost, has changed the basic paradigms that had previously been thought to be valid. The nightmare of Malthusian collapse of human society due to the failure of agricultural productivity to keep pace with population growth is being increasingly predicted. The UN Food and Agriculture Organisation (FAO) projects that global agricultural production will need to more than double by 2050 to close the gap between food supply and demand. However it is not just the amount of food that will need to be a focus. Livestock takes up almost 80% of global agricultural land and global meat consumption is predicted to rise both on a per capita and an overall level. At the moment the world population uses about 50% of the habitable land for agriculture. If the world were to adopt the national diets of different countries then the ability of the global land mass to feed the world population varies. So, for example, if the whole global community adopted the dietary habits of China or India then, currently, we would actually need less land to meet this requirement. Global adoption of a UK diet would require some expansion of agricultural land however this is within the constraints of the land that is available. If the dietary habits of some countries were to be adopted globally (e.g., North America, Brazil, Argentina, Australia and several countries across Europe) then it would be impossible to produce enough food. For example, if the whole world was to adopt the dietary habits of the USA and we converted all of our habitable land to agriculture then we would be almost 40% short of land to grow plants and livestock on. Lloyds of London produced a report in 2015 in which they modeled the effects of a shock to global food production. They hypothesised that "Sudden disruptions to the supply chain could reduce the global food supply and trigger a spike in food prices, leading to substantial knock-on effects for businesses and societies. The food system's existing vulnerability to systemic shocks is being exacerbated by factors such as climate change, water stress, ongoing globalisation, and heightening political instability." Essentially they expressed concern about the lack of robustness in the system and the resilience of the global food supply. So the first major challenge is to try to ensure the food supply and to build greater robustness into the system.

The second great challenge relates to over consumption. While the global food supply lacks sufficient resilience, societies throughout the developed world are reaping the consequences of overweight and obesity. Chronic diseases such as type 2 diabetes, circulatory disorders (cardio and cerebro vascular disease), arthritis and other muscoskeletal disorders and diseases caused by local and systemic chronic inflammatory conditions are all exacerbated by the consequences of consumption of too much (and in some cases the wrong type) of food. As lifespans increase the consequences of this diet-related disease will continue to cost significant amounts of money and other resources. Changing dietary habits is difficult – particularly when the consequences of those changes are not necessarily immediately visible.

The third great challenge is linked to chronic health issues. Despite considerable research efforts over the past 100 years or more, and the increase in knowledge about non-nutritional (bioactive) components of foods including plant compounds, pro and pre biotics and other sources of potential benefits including algae and yeasts, the use of food as a medicine is still remarkably poorly exploited. One of the problems has been the use of a pharmaceutical type model which relies upon the precepts of a 'magic bullet' developed by Ehrlich. The principle built upon an old superstition that bullets could be 'charmed' so as to hit only one particular individual. So Ehrlich searched for compounds that could destroy a causative agent (the bacterium that caused syphilis) and not affect the human host. This led to the discovery of Salversan. This success encouraged a school of German physiologists, including Emil von Behring, to propose that researchers should develop specific drugs to target specific germs, attacking the cause of the disease directly, rather than treating the symptoms. Much of the research around bioactives in food is still around treatment of symptoms since the effects of food chemicals is almost invariably long term and predictive tools are not generally available.

The fourth great challenge is closely linked to the resilience of the food supply. Food waste, the circular economy and more efficient use of resources is a major challenge for the agri-food industry. There are opportunities all along the supply chain to develop better practices and to use technology to make food production more efficient. One of the low hanging fruit in this area is a reduction in animal protein consumption in developed countries. In terms of efficiency of conversion, animal protein is considerably more wasteful than plant protein. However plant protein is not necessarily as easy to process using current technology and personal preference for animal protein is significant. However the challenge extends to food preparation and packaging as well as the food itself.

The final challenge is to ensure that any solutions that are developed win the support of consumers. We are all food experts in the sense that we are consumers and are highly critical of what we consume. Underpinning consumer confidence is confidence in the food supply, in legislation and regulation to ensure food is safe and of high quality. These assurances must be in place if we are to continue to have a strong and viable food system which meets the challenges facing us.

OPPORTUNITIES AND SOLUTIONS

The key opportunities and solutions can be summarised in terms of the simplified diagram below. The primary chain goes from the farmer/producer to the processor, the food producer through the retailer to the consumer. The first and obvious opportunity is to recognise the strain points in the primary chain and to ensure that the highest value part of the chain (generally the consumer), feeds back along to the other parts. So the feedback loops connect all parts of the chain. However the *flow* of the chain is essentially linear and no amount of trying to engineer responsiveness will change that basic reality. There is, however, one definite cut off point which will, for any product, completely halt the chain. If the consumer does not want the product then the chain ceases to function. The further back in the chain, the less room there is for manoeuvre. If the product on the shelf is not selling because of the way it looks then it can be repackaged. If taste and flavour are a problem then reformulation may be the answer. However, if the consumer does not want the product because of the raw materials then the whole chain collapses. This may turn out to be the case with introducing insects into the food chain. It is also the case for some groups of the population who will not consume certain foods e.g., vegetarians. Repackaging or reformulating meat will not make it more appealing to a vegetarian or a vegan. So the challenge and opportunity is to try to ensure that products are able to be reformulated, repackaged and reprocessed quickly and this means adopting the prototyping approach of additive manufacturing.

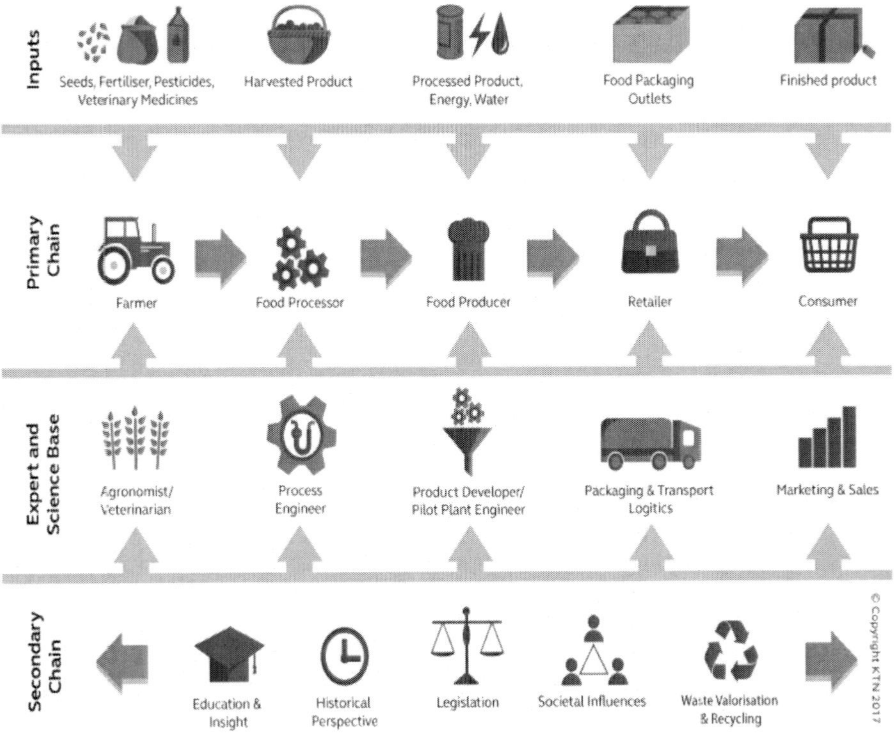

Figure 1: Key opportunities and solutions in the food supply chain.

Moving on from the primary chain, each part of the chain has specific inputs. Each of these must be optimised and each can have an effect on the final product and on profitability. The inputs can be considered as costs. If the product from each part of the chain does not exceed in value the cost of the inputs then the chain will collapse. That is why raw material costs can be so crucial and why there is so much pressure on farmers. Each successive link in the chain will try to drive cost out of their part of the chain. For many it is reducing the cost of the preceding part of the chain. For some it means driving down their own costs as well. So retailers will try to ensure that the product they receive from the producers is a cheap as possible so that it is attractive to consumers. They will also try to drive down packaging costs and this involves a whole different value chain. However if retailers reduce costs from packaging (or storage or logistics) then this saving is not passed back down the chain to the producer, processor or farmer. It is passed up the chain to the consumer or to the retailers' shareholders. This means that the producers must also drive their costs down in a similar way as do the processors and, ultimately, the farmers. The farmers, however, can only reduce their own costs since they are at the start of the chain. There are, then opportunities for all parts of the chain to reduce costs but also to add value to their products. This can be through the use of additive manufacturing or other techniques which can both add value and reduce costs or to produce foods that have greater value than simple nutrition. This can be through using health and wellness as a prime purchasing driver, for example. Of course there are some examples of intrinsically high value ingredients (e.g., saffron) and in this case the primary producer is in a strong position due to the uniqueness of their product. However there are relatively few examples of this and the industrialisation of agriculture together with improved transport infrastructure may threaten this type of value proposition. For example asparagus used to be expensive however because of trade agreements, transport, the development of new varieties etc., the cost has fallen in real terms. However it remains a viable crop since, while the margins have dropped, the volumes have increased to maintain profitability for the grower.

In addition to physical inputs, the food chain also benefits from the expert and science base. These are also areas where solutions can be found and new opportunities created. In many cases the inputs are obvious and direct. For example, the application of agronomy to the husbandry of crops or the input of process engineers into food processing. However there are also some interesting opportunities for cross-sectoral interactions. For example, the development of autonomous vehicles could have an impact on harvesting, transport and logistics. Product development is strongly influenced by the final product as well as by the raw material. So, for example, if the final product contains specific properties related to health then these have to be confirmed and proven and so clinical scientists and investigators of biomarkers are often involved. Once an effect has been demonstrated then crops can be grown to optimise the presence of the active principles, processed so as to preserve that activity then food produced with optimal bioavailability. All along this process, the taste and sensory characteristics of the food have to remain optimal. Other examples include the use of satellite systems for improved crop protection as part of a management system and the development of robotics, machine learning and robotics in processing plants. The psychology of food choice also relies heavily on neurology and understanding of consciousness using mathematical formula such as integrated information theory. Such a theory can also look at societal consciousness and this can include bacteria in the gut microbiome and other ecosystems. All of these areas can attract funding and can improve various aspects of the primary chain.

This brings us to the secondary chain. While we have acknowledged that the use of the expert and science base is not only linear but is also cross sectoral, there is also a further set of influences that apply across the whole food chain. These components of the secondary chain include education and awareness, historical perspective (what are people accustomed to eating, what is their perception of certain foods), what is the legislative position, what are the societal influences and how can we improve efficiency by incorporating waste valorisation and recycling all along the food chain. Some of these provide opportunities for chemists and other scientists to develop new and better systems however in some cases conformity to accepted norms is more important if a food is to successfully reach the consumer. The other opportunity is for chemistry to provide support for these components of the secondary chain. This can include analysis in support of legislation and regulation, tests of quality and authenticity and educating people about the science underpinning foods so they can make an informed choice.

Science in general and chemistry in particular are key to meeting the challenges of the future of food and to improving the health and wellness of the population at an individual and societal level.

The five great challenges outlined above will need different solutions however chemistry lies at the core of each of them as the ultimate enabling technology.

The key driver is the interconnectiveness of food systems. Chemistry and other disciplines will impact on how plants and animals are grown, how to make that process more efficient, how to drive health and wellness by growing and processing food, how to develop new methods of manufacture and production by using lessons learned from the bioeconomy and innovations like additive manufacturing. Underpinning this must be the establishment of a deeper understanding of the molecular, physical and macromolecular properties of food.

In addition, industry must be able to adopt new technologies and use them to prepare new and better products on a scale that will meet demand, consumers must have confidence in the food that is developed, in the safety and quality of food and its ingredients, in the science and technology which is being used and they must be able to afford to buy healthy, safe and nutritious food. Legislators and regulators must have an understanding of the benefits of technology and how best to ensure that innovation is encouraged and brought to fruition.

In the following chapters, experienced experts have given greater details of the underlying challenges in each of the areas and have suggested solutions and steps that can and should be taken to meet these significant challenges. This includes assuring a robust food supply system (Chapter 1 and 6) through primary production (Chapter 2) and processing (Chapters 8,10 and 11) to deliver safe (Chapters 4 and 5), healthy (Chapter 7) and acceptable food (Chapters 3, 9 and 12).

The scientific, socioeconomic and practical challenges are also great opportunities to develop a resilient agri-food industry built upon a cohesive, interlinked and cohesive supply chain all the way from primary production to the consumer. Such a chain is not a linear entity. Rather it has feedback loops and mutually beneficial interactions. Food has progressed from simply being a means of providing sustenance to being an integral part of life. It has the potential to change lives for the better, to help in the fight against chronic disease, to overcome natural and man-made challenges and to be the spur for new science and technology. Chemistry is the key underpinning technology that will enable this to happen.

ACKNOWLEDGMENTS

The authors would like to thank Pat Bain from the University of Aberdeen Rowett Institute for the original artwork and the RSC Food Group for supporting this publication. Wendy Russell and Madalina Neacsu would like to acknowledge funding from the Scottish Government Strategic Research Programme.

The authors would also like to thank the Food Sector Group of the Knowledge Transfer Network for their initiation of and support for this project.

In: Food: The Vital Ingredient ISBN: 978-1-53614-402-4
Editors: A. B. Hanley and W. R. Russell

Chapter 1

GLOBAL FOOD ISSUES AND THE ROLE OF TECHNOLOGY

***A. Bryan Hanley*[1] *and Wendy R. Russell*[2]**
[1]The Knowledge Transfer Network, London, UK
[2]University of Aberdeen Rowett Institute,
Aberdeen, Scotland, UK

INTRODUCTION

Food and agriculture are global businesses. In addition to having international reach, they are also users of technology and advanced methods, materials and processes. Growing, processing and transporting food has to be underpinned by fundamental disciplines and discoveries in physics, chemistry and biology as well as transport, storage, logistics, packaging and many others. In addition to the traditionally accepted skills bases, there are a plethora of new technologies adapted from other areas. For example, precision agriculture makes use of advanced artificial intelligence global positioning systems. In future the development of autonomous vehicles is likely to offer great opportunities in growing and harvesting crops as well as in transporting raw materials and finished products. Pressure on the use of pesticides and other plant protection products has encouraged research on insect physiology so as to enable the development of precisely targeted products for pests that have no effect on beneficial insects. Greater knowledge of taste receptors, GCPRs, neurology, psychology and sensory assessment underpinned by advances in machine learning, sensor technology and neural processing will continue to impact upon food choice, consumption and liking. In many cases the crucial interaction between molecules and chemistry is the key discipline that underlies advances in these areas.

The primary drivers for food production were originally ease of growth, harvest and yield. This led to the adoption and spread of best practice including the use of particular crops and the development of varieties that grew well and produced sufficient yield to sustain the community. However, apart from agronomic advances such as crop rotation, yields remained comparatively low until the 19th century.

A major breakthrough in the partial industrialisation of agriculture came with the development of the Haber-Bosch process. The reaction is as follows.

$$N_2 + 3H_2 \rightarrow 2NH_3$$

This apparently simple process could only be successful by the use of a metal catalyst. Both Haber (who developed the process) and Bosch (who devised the scale up) were awarded Nobel prizes for Chemistry. The Haber process now produces almost 500 million tons of nitrogenous fertiliser a year.

Another key advance in crop yields was the development of pesticides and other plant protection products [1]. Around 4,500 years ago the Sumerians used sulphur compounds to control mites and other pests. Smoking was a common method to prevent fungal disease such as mildew and pyrethrums from the dried flowers of *Chrysanthemum cinerariaefolium* have been used for over 2,000 years. The major explosion in pesticide use came in the 1940s with the development of DDT and other related compounds. This greatly supported the industrial scale agricultural production of food. In the 1970s and 1980s the world's bestselling pesticide, glyphosate, was introduced. More recently the use of genetic modification to introduce plant protection products directly into the plants was developed. All of these rely on knowledge of the chemistry of the systems.

In the case of food processing the major traditional technologies are refrigeration and freezing, canning, irradiation, dehydration, freeze-drying, pickling, pasteurizing and fermentation. More recently the development of techniques such as additive manufacturing (3D printing) have shown some promise but, so far, little commercial reality [2]. However what is clear is that chemistry will continue to have a major role to play in the future developments of food and agriculture.

Indeed it is interesting to review the progress of food and agriculture from the standpoint of chemical advances. In many cases, without the underpinning chemistry, the advances in food and agriculture would not have been possible.

Some of the key Nobel Prize winners and their relationship to food and agriculture are listed in Table 1.

AGRICULTURE AND PRIMARY PRODUCTION

It is clear that chemistry has been implicated – directly and indirectly, in many advances in agricultural and food science. From understanding the basic reactions and properties of food systems to nutrition and food chemistry and the role of individual chemicals and their interactions, the development of understanding of foods through their inherent chemical properties has been crucial. The challenges and opportunities outlined elsewhere in this book will also require the application of chemistry in new areas and new environments.

Global food production has increased substantially and has largely kept pace with, or exceeded global population growth in the past. The Malthusian disaster has not happened.

Table 1. Nobel Prize winners and their relationship to food and agriculture

Prize-winner and Year	Citation	Link to Agriculture and Food
Fischer 1902	"in recognition of the extraordinary services he has rendered by his work on sugar and purine syntheses"	Knowledge of purines in plants and animals and their importance as building blocks. First synthesis of a protein and attempt to produce synthetic food.
Buchner 1907	"for his biochemical researches and his discovery of cell-free fermentation"	Cell free fermentation of sugars which is the basis of food fermentative technology. Buchner was also a professor at the University of Berlin.
Wilstatter 1915	"for his researches on plant pigments, especially chlorophyll"	Showed that chlorophyll is a mixture of two compounds and independently invented paper chromatography.
Windaus 1928	"for the services rendered through his research into the constitution of the sterols and their connection with the vitamins"	He described the structure of the plant cell component ergosterol, and explained the structure and function of vitamin D,
Hardin and Von Euler-Chelpin 1929	"for their investigations on the fermentation of sugar and fermentative enzymes"	Hardin worked on the antiscorbutic and anti-neuritic vitamins and their occurrence in food.
Fischer 1930	"for his researches into the constitution of haemin and chlorophyll and especially for his synthesis of haemin"	His expertise in plant and animal pigments, including chlorophyll and haemin, greatly increased understanding of these fundamental building blocks.
Karrer 1937	"for his investigations on carotenoids, flavins and vitamins A and B2"	Karrer established the correct formula of vitamin A and confirmed the structure of ascorbic acid.
Haworth 1937	"for his investigations on carbohydrates and vitamin C"	Haworth carried out the first synthesis of ascorbic acid and the Haworth Projection' which is a 2-dimensional representation of 3-dimensional sugar structures, is still used.

Prize-winner and Year	Citation	Link to Agriculture and Food
Ruzicka 1939	"for his work on polymethylenes and higher terpenes"	Ruzika completed the first synthesis of b-collidine and of linalool and the partial synthesis of pinene.
Virtanen 1945	"for his research and inventions in agricultural and nutrition chemistry, especially for his fodder preservation method"	In addition to his work on agriculture and nutrition, Virtanen was able to show that phosphorylation is the first step in many fermentation reactions. He also carried out the isolation of many mustard oils (isothiocyanates) and first isolated the indole glucosinolates, glucobrassicin and neoglucobrassicin.
Robinson 1947	"for his investigations on plant products of biological importance, especially the alkaloids"	Worked on plant anthocyanins which at the time were used as dyestuffs and as natural food colourants. They are now considered to have health promoting properties.
Calvin 1961	"for his research on the carbon dioxide assimilation in plants"	His parents had a grocery store and he worked on Chlorella which has been suggested to contain compounds that improve the lipid profile in humans
LeLoir 1970	"for his discovery of sugar nucleotides and their role in the biosynthesis of carbohydrates"	Invented a sauce called Salsa Golf comprising ketchup and mayonnaise. Commonly called Marie Rose sauce.
Diesenhofer, Huber and Michael 1988	"for the determination of the three-dimensional structure of a photosynthetic reaction centre"	Diesenhofer's family owned a farm – which he showed little interest in. Huber worked on the structure of trypsin inhibitors – plants such as soya contain trypsin inhibitors and this is in part responsible for their anti-nutritional properties.
Stoddart 2016	"for the design and synthesis of molecular machines"	He has published on the potential of edible MOFs and on their use

Prior to 1800, when the global population was 1bn, the annual population growth rate was < 1%. Between 1900 and 2000 there was a period of accelerated growth and the population rose from 1.5bn to 6.1bn.

While the growth rate has slowed again, current expectations are that the global population will reach 9.7bn by 2050.

Since 1800, agricultural production has outstripped population increase. In addition, average caloric consumption was lower in 1800 than it is today and larger proportions of the population had a nutritionally inadequate diet [3]. The FAO have estimated that gross agricultural output increased by 60% from 1938 to the late 1950s and doubled between that time and 2001. Agricultural production was slow in the 18th century but began to accelerate in the 19th century [3]. There was a 70% increase in agricultural production in England and Wales between 1801 and 1851 due to increased yields and the spread of new and more efficient agricultural practices. So, in the case of wheat, the yield during the middle ages was 10 bushels per acre throughout most of England. Norfolk was an exception where yields of perhaps 20 bushels were sometimes obtained. These high yields were achieved by sowing twice as much seed as elsewhere; Norfolk farmers, in other words, had not pushed the yield-seed ratio above the standard of other counties. Around 1800, several large-scale surveys show that the yield of wheat was 20-22 bushels per acre across England. Yields had doubled in most places. For example, average corn yields in the United States remained relatively flat throughout the 1800s until the 1930s. In the period since 1940, yields have increased more than five-fold. The process of increasing yield has largely halted and global yields of major crops are now stagnant.

Changes to eating habits and the increase in the population means that agricultural production may need to increase by an estimated 70 per cent globally and by 100 per cent in developing countries by 2050 [4]. About 12% of the land area of the planet is used for agricultural production and there is little prospect of that figure being increased in the future. Increasing urbanization, loss of farmland through global warming and the unsuitability of much of the terrain for agricultural use means that pressure ofn food supplies will require the development of new technologies to grow, process and more efficiently use food. This technological challenge applies across the whole food chain from primary production to the consumer and even beyond that to the issue of waste and the competition for resources such as water and energy and environmental impact.

Some of the proposed solutions are unlikely to have the desired effect. For example, trying to find alternative non-plant protein sources such as insects will only ever be of limited value. While insects can grow and thrive on what is generally considered to be non-utilisable waste, the conversion factors for insects is similar to that for chickens. It has been found that, in a like for like trial, insects (crickets) fed a poultry-feed diet showed little improvement in protein conversion efficiency compared to chickens. The best conversion rates were obtained from those crickets given a high quality feed so they will be competing for feed with other commercial animals (cattle, pigs and sheep). In addition it has been noted that the protein content of insects has been consistently overestimated due to the presence of other sources of N in insects. The protein content is usually calculated from total nitrogen using the nitrogen-to-protein conversion factor (Kp) of 6.25. This factor overestimates the protein content, due to the presence of nonprotein nitrogen in insects. It has been found that a specific Kp of 4.76 ± 0.09 was calculated for larvae from Tenebrio molitor, Alphitobius diaperinus, and Hermetia illucens, using amino acid analysis. After protein extraction and purification, a Kp factor of 5.60 ± 0.39 was found for the larvae of three insect species studied. Even the adoption of changes in the regulatory regime in the EU to permit the feeding of insects to fish comes cloaked in caveats.

Current regulation still bans the use of unprocessed former food (meat and fish), catering waste and slaughterhouse products and animal manure. The permitted feedstocks for the insect larvae are unprocessed former food (dairy and eggs) and vegetal substrates. In addition, since insects were not widely consumed in the EU before March 1997, insect-based foods are considered 'Novel Foods'. As such they fall under Regulation 2015/2283 and require authorisation from the European Food Safety Authority (EFSA) before being placed on the EU market. However farmed fish require a source of long chain polyunsaturated fatty acids (LC-PUFAs) such as Docosahexadienoiic acid (DHA) since fish are unable to produce these 'healthy oils' themselves. Using fish derived fish meal or adding back LC-PUFAs from algal sources has been the normal approach. The increased cost of LC-PUFAs has increasingly made this uneconomic and as a result the levels of LC-PUFAs in farmed Scottish salmon halved between 2005 and 2015. One approach which might be timely is to genetically modify insect larvae so that they can produce LC-PUFAs. A second possibility would be to incorporate LC-PUFAs into the diets of insects and to see whether they could accumulate these preferentially. In either circumstance, a fish feed containing these would command a premium price.

More efficient use of resources is a key factor in enabling primary production to continue to provide sufficient protein, fats and carbohydrates to feed the growing population.

PROCESSING

The processing of raw materials into foods has been one of the most significant success stories of the past few hundred years. Preservation and processing have enabled humans to break free from the famine:feast cycle that was so much part of earlier life. Hunter-gatherers had periods of significant food intake interspersed with times when food was scarce. More recently the effects of famine during pregnancy in modern times were found to be significant for the maturation of the pre-natal infant. Such a pattern of deprivation leads to 'thrifty baby syndrome' and a more efficient utilization of resources in the adult [5]. This might suggest that having a first trimester during January to April might equip the growing fetus with the programming to accumulate nutrients. This may have been of value for hunter gatherers but is less advantageous when food is generally plentiful.

The epidemiologist David Barker was one of the first to try to quantify what undernutrition in utero might lead to in later life. Lucas coined the term fetal programming and opened the field for extensive research into the fetal origin of adult diseases. The association between low birth weight which reflects intrauterine nutritional status, and the development of adult diseases has been confirmed in many studies for type 2 diabetes, hypertension and ischemic heart diseases. Since the average life expectancy of the Paleolithic hunter gatherers was perhaps between 20 and 30 years, long term disease issues such as hypertension and type 2 diabetes are unlikely to have had a significant effect.

Even later when agriculture had become established, spring was typically a hungry period. Before the crops began their new cycle, people had to largely live off what they had harvested in the autumn and managed to store over the winter. Even the types of crops that were grown had an effect on the efficiency of their storage. Those with high water content like potatoes, fruits and some vegetables were difficult to store. They tended to succumb to bacterial and/or fungal infection and were therefore much more liable to rot in storage. The development of pickling and preserving helped to ensure that at least a part of the food supply would be preserved. Grain crops, which have a much lower water content, could be stored more successfully – if they were dried properly before storage. In addition, bread and other foods produced from grain could be stored more successfully that raw potatoes and also provided some dietary variety.

Many early food processing techniques were firmly aimed at preventing spoilage. Drying (either in the sun or over a fire) were (and still are) used to preserve some high water crops such as tomatoes. At the other extreme of temperature, natural freezing was used along with cool environments such as caves and even streams. These are both essentially natural events where conditions serve to prevent the growth of spoilage bacteria and fungi. Fermentation lies further along the path. Although it is a natural process, the growth of yeasts on food can also spoil it so beginning to understand how fruits can be converted into wine or how conditions to make acid fermented foods such as sauerkraut can be developed was a significant conceptual leap. Pickling is essentially akin to acid fermentation without the fermentation step. It prevents bacterial growth and preserves the properties of the food. Both curing and making preserves are similar in that salt and sugar respectively are used to lower the water activity of the food and to thereby prevent microbial growth.

More modern techniques including canning, pasteurization, irradiation, the addition of chemical preservatives, modified atmosphere, high pressure and bio preservation have all helped to increase storage times, prevent microbial growth and preserve as much of the initial characteristics of the processed food as possible. However modern food processing is also designed to alter other characteristics of food such as ease of cooking and taste and texture characteristics. Some of these were technology led. For example the development of some cooking techniques relied on the introduction of stable, non-flame, indirect heating. Microwave ovens, which have led to the rapid acceleration in the development and use of convenience foods, were a significant breakthrough in kitchen as well as in industrial technology. Extrusion led to the range of products and allowed control of food shapes, constituents and other characteristics.

The future of food processing may in future be driven by alternative priorities. The development of additive manufacturing may enable the production of feedstock blind foods. As protein sources become less available, the ability of food producers to create food from a range of starting materials without having to produce a separate piece of equipment (or factory) tailored to a particular set of ingredients may become imperative. In addition, health drivers may continue to provide challenges for food manufacturers to produce food containing less carbohydrate and salt and lower energy density over a range of food groups. If the price or protein rises and availability becomes less, then a better understanding of protein uptake by humans and how to optimize the most effective utilization of protein may become increasingly important. Finally, despite many years of study and investigation, there is still a very incomplete understanding of the microbiome and how to manipulate it and of the potential of plant phytochemicals to influence human health and behavior. Both of these areas require significant investment if we are to reduce the burden of chronic ill health and build a better and more responsive agri-food system.

This will in turn, lead to a better understanding of the optimization of food intake and therefore food processing to meet this end, rather than the maximization of food consumption.

Global Consequences

The global economy is still strongly reliant upon a robust and resilient agri-food network. Trade between continents has allowed Europe to operate as a global market at any time of the year and the subsequent loss of seasonality of produce allows us to buy what we want, pretty much whenever we want. As supply chains have become geographically longer it is increasingly difficult to have confidence in the fidelity of products. Partial and pre-processing is carried out wherever it is convenient to do so. This has led, in some places, to a backlash and to a rediscovery of localism and the authenticity (and hence provenance) of foods and ingredients that are part of them. However this inevitably increases costs and there is always a danger of unscrupulous practices being introduced. It is therefore essential that chemistry keeps pace with potential hazards which could be introduced and to ensure that people are actually buying what they think they are.

The discovery of new resources, a better understanding of how food is used, methods to reduce, refine and repurpose waste new technology to create value for the whole supply chain and maintaining the confidence of consumers is all crucial to the future of the global agri-food system.

REFERENCES

[1] Kogan, M. 1998 Integrated pest management: historical perspectives and contemporary developments. *Annu. Rev. Entomol.*, 43:243 - 70.

[2] Lipton, J. I. Cutler, M., Nigl, F., Cohen, D. and Lipson, H. 2015 Additive manufacturing for the food industry. *Trends in Food Science & Technology,* 43/1 114 - 123

[3] *The World Bank Health, Nutrition and Population dashboard* http://datatopics.worldbank.org/health/population Accessed 2017.

[4] *How to feed the World in 2050 FAO Report Rome 2009* http://www.fao.org/fileadmin/templates/wsfs/docs/expert_paper/How_to_Feed_the_World_in_2050.pdf.

[5] Negrato, C. A., Gomes, M. B. 2013 Low birth weight: causes and consequences. *Diabetol. Metab. Syndr.*, 2013 2; 5:49.

In: Food: The Vital Ingredient ISBN: 978-1-53614-402-4
Editors: A. B. Hanley and W. R. Russell

Chapter 2

TECHNOLOGIES SUPPORTING RESILIENCE IN PRIMARY AGRICULTURE

Wendy R. Russell[1] and A. Bryan Hanley[2]
[1] University of Aberdeen Rowett Institute, Aberdeen, Scotland, UK
[2]The Knowledge Transfer Network, London, UK

INTRODUCTION

Primary agriculture underpins the whole food chain and is the start point for food security, and for nutrition and health. It is a key driver for the global economy and impacts directly on livelihoods and the environment. According to the World Food Bank 'agricultural development is one of the most powerful tools to end extreme poverty, boost shared prosperity and feed a projected 9.7 billion people by 2050 [1]. For it to be effective, agricultural practice needs to be both sustainable and visionary, as it will be the major contributor to solving our current and future societal and environmental challenges.

Although a new and evolving concept, sustainable agriculture can be defined as a farming system that provides resilience in the food supply chain, with scope to resist transient perturbations upset. It provides economic growth for reinvestment and development of new and better technologies to ensure that food stocks and renewable energy are available for future generations. It conserves biodiversity, while minimising degradation and loss of habitat and ensures that utilisation of resources is not conducted at a rate where they are irreversibly depleted and it ensures that CO_2 emissions are balanced by fixation. Economic growth is also dependant on agriculture. In 2014, it accounted for one-third of global gross-domestic product [2] and approximately one billion people were employed in the agricultural sector [3]. The current food supply system is at risk from a number of directions and this threatens not only human health, but also the global environment.

With increasing population there is competition for natural resources and humans are creating unprecedented levels of pollution and waste. Unfortunately this use of primary agriculture and the related food chain does not result from societies that are replete in terms of the share of global resources. Approximately three billion people are undernourished or eating the wrong types of food and at least another two billion people are overweight or obese [4]. Both of these extremes put pressure on health systems and are an economic drain on countries. Uncertainties with regard to trade, societal and climate change make addressing these challenges even more complex, but it is clear that agriculture practice needs to adapt to address both local and global burdens.

COMPETITION FOR NATURAL RESOURCES

Ecosystem dysfunction is a direct consequence of inefficient land management and the FAO estimates that approx. 1.5 million people are dependent on degraded land. This includes deforestation, desertation, soil erosion, mineral depletion and chemical degradation (acidification and salinization) [5]. Agriculture withdraws more than 70% of freshwater resources and depletes underground aquifiers at an unsustainable rate.

Several areas, particularly northern China, Western US and the Upper Ganges are already severely depleted with Iran, Mexico and Saudi Arabia also at risk. Increased livestock production also impacts on native wildlife species and a range of ecosystem processes [6]. Agricultural productivity increased dramatically in the latter part of the 1900s due to energy intensive practices and increased use of pesticides and fertilisers. There was a fourfold increase in yield between from 1945 to the end of the century [7]. Much of the energy needed is fossil-fuel derived and this reliance on petrochemicals is a concern for future agriculture costs and concomitant food shortages. Although direct and indirect on-farm costs are around 2% (UK and US) of total energy use, off-farm costs of food production (processing, packaging, manufacturing, transport, waste disposal) can be 5-7 times this amount.

Changes to our climate has immediate impact on crop yields and this can be especially burdensome in food-insecure areas such as sub-Saharan and eastern Africa and areas within south and east Asia [5]. Changes in marine-food production is also expected with movement of aquatic species into new waters with negative consequences for some fish stocks [8]. According to the FAO report Agricultural emissions (crop and livestock) increased from 4.7 billion tonnes of carbon dioxide equivalents in 2001 to over 5.3 billion tonnes in 2011 [5]. Improvements in practice need to facilitate a movement from agriculture being a contributor towards, to being beneficial in mitigating against climate change. Key to this will be the development of climate smart technologies and practices. Climate smart agriculture needs to be adapted to different site-specific conditions.

Strategies to use natural climate solutions (NCS) to both increase carbon storage and reduce greenhouse gas emissions are being considered. These include conservation/restoration projects and improved land management. It was demonstrated that NCS could possibly provide more than one-third of the mitigation required to stabilize warming to below 2°C (and meet the Paris Climate Agreement) by 2030 and also potential to improve soil productivity, clean air and water and maintaining biodiversity [9]. GHG emissions could also be reduced through improved use of fertilisers. Overuse of synthetic nitrogen-based fertiliser has significantly contributed to GHG emissions. It is estimated that in China (the largest producer) nitrogen-based fertilizer-related emissions result in approximately 7% of total GHG emissions from China and that this could be cut by 2 - 6% by technological advances in fertiliser use [10]. With regard to rice production, alternate wet and dry irrigation compared to continuous flooding in certain climate conditions can significantly reduce GHG emissions [11]. Ground cover rice production systems also reduced water use and GHG emissions [12]. Other technological improvements include waste management and improved livestock breeding (see below).

LIVESTOCK PRODUCTION

Livestock product consumption various between countries and economic status, but globally contributes on average to approximately 17% of kilocalorie consumption and 33% of protein intake [13]. It comprises milk (619 million tonnes), meat (259 million tonnes), fish (130 million tonnes) and eggs (63 million tonnes). Meat production tripled from 1980-2002 [14], but this was predominantly driven by the countries in transition, as in higher economy countries the growth is much slower, albeit it starts from a high level. Increases in meat consumption are likely to be driven by a combination of increasing population, urbanisation, expectation and income.

Increased production of livestock is accompanied by changes in land use. This includes grassland (shrubland, rangeland, and pasture) and mixed methods of production, as well as high protein crop production for animal feed. In particular soya production has expanded considerably in South America, with transition countries doubling (to 36%) in their share of production of cereals for animal feed [15]. Decades of research have improved our understanding of the nutritional requirements of farm animals, yet poor nutrition is still a barrier to production, particularly for smallholder farmers. Investment in technologies both in data capture (growth, body composition, inputs and outputs) and modelling these systems will improve production [16]. It will be essential that we access sustainable plant based protein sources, including potential aquatic sources, and microalgae, fungi and insects. Insect protein was shown to influence the gut microbiota of laying hens, facilitating the potential to degrade the chitin and an increased production of short chain fatty acids in laying hens [17]. Advances in genome editing (e.g., CRISPR/Cas9) could advance the genetic and biological engineering of microalgae as a valuable feedstock [18].

Growth-promoting steroids and steroid-like compounds (GPSCs) have been use for decades to improve growth, feed efficiency and increase lean muscle mass, which in turn results in lower meat process for the consumer [19]. There are concerns regarding the environmental impact of growth hormones, in particular their presence in surface and groundwater [20]. Better understanding regarding their *in vivo* metabolism, presence in food and environmental samples, as well as use of alternatives (e.g., regulation of gut fermentation) will inform both application and legislation of these products.

Breeding has been responsible in part for the increased in yield of global livestock products and this has been through predominantly conventional techniques, such as breed substitution, cross-breeding and within-breed selection [21]. Selection within breeds of pig, poultry and dairy cattle produces genetic changes in the range of 1 – 3% yield per year, but these are often lower for sheep and beef cattle [22]. Future breeding strategies will include better DNA-based technology, which although already valuable, will be utilised more extensively for meat production and could inform on wider traits including disease resistance, fertility and environmental aspects including methane production. Transgenic livestock is also now feasible, and publication of the cattle genome sequence is likely to facilitate improvements in both milk and meat production [23]. Animal demonstrating agriculturally useful traits will be continually sequenced, but this must be on balance with preservation of genetic resources of our livestock, which will be critical in adapting to meet future food challenges.

Aquaculture (fish, crustaceans and molluscs) is currently the fastest growing sector globally at 47.1 million tonnes and 26.7 million tonnes for inland and marine production respectively in 2014 [24]. Capture fishing has increased relatively moderately from 90.2 million tonnes in 2009 to 93.4 million tonnes in 2014. Whereas aquaculture increased from 55.7-73.8 million tonnes in the same period, with China responsible for more than 60%

of global production (FAO 2016). With this rapid expansion comes a need to optimise feed production especially provision of species-specific feeds that meet the nutritional requirements of different life stages of the farmed species [25].

Zoonotic diseases (both endemic and emerging) and an important consideration for agriculture and is critical for maintaining global health security. Brucellosis, rabies, zoonotic influenza, anthrax, haemorrhagic fever and salmonellosis are among the top ranking prioritized zoonoses [26]. In addition to building capacities and strengthening partnerships for these diseases, effective detection and development of treatments for emerging diseases and affordable and improved treatment for existing conditions will facilitate an improved response to this world-wide threat.

Agriculture has and will continue to play a significant role in the worsening antimicrobial resistance (AMR) that is now a major threat to animal and human welfare [27]. Veterinary and human medicine relies extensively on antibiotics to treat bacterial infections and this, in part, has contributed to resistance towards available antibacterial agents. However, the wide-spread global use of low-dose antibiotics as growth-promoting agents in intensive agricultural practices was considered to have promoted antibiotic-resistant enteric bacteria [28]. There is a need to meet food requirements without continuing to contaminate the environment antibiotic-resistant microbes and genes [29]. Improving livestock production though alternative methods are key. Antibiotic use is less intensive in organic production [30]. Additionally, decreasing global meat consumption will eventually play a role, but immediate improvements in sanitation and hygiene are required.

AGRONOMY

Global scale agrimony is now a major topic, whereas previous practice was more generally focused on a local level [31]. With impacts from global warming, agronomic control of nutrient availability and flow is crucial. It is frequently cited that there is a need for 70 – 100% more food in the next few decades [32] and agronomic practice will have to change. There is much debate as to whether smallholder farmer-dominated or large-scale intensive farming will deliver efficiently in terms of both production and maintaining biodiversity [33]. Achieving agronomic productivity, while maintaining biodiversity is a significant and will be a continuing global challenge [34]. Increased global trade has also resulted in an increase in invasive species [35]. It is estimated that 11000 alien species have invaded Europe and 30% of them have caused economic damage or harm to biological diversity [36]. With climate change, this is likely to rise [37]. Important topics from a technologies perspective are air, soil and water quality crucial role of and the importance of nutrient management within these systems [38, 39] including the impact of nitrogen fertilization on greenhouse gas emissions [40].

Plant science will play a major role in addressing food security in terms increasing productivity, nutrient quality and efficiency in use of resources [41]. Crop yield increased dramatically from the latter part of the 1900s, predominantly as a consequence of new technologies such as breeding, fertilisers and fungicides [42]. Although development of these practices and emerging new technologies are likely to continue to increase yield, it is recognised that to ensure food and nutritional security will require much consideration of the overarching primary agricultural practice and the wider food supply chain.

Innovative farming methods will be required to meet the expansion of urban landscapes and will include vertical and aquatic farming of plants [43]. In 2014, production of farmed aquatic plants was approximately 27,307 thousand tonnes [24]. The majority are seaweeds and Indonesia the largest producer. The most prominent species are: Kappaphycus alvarezii and Eucheuma spp. (10,992 thousand tonnes), Laminaria japonica (7,655 thousand tonnes), Gracilaria spp (3,752 thousand tonnes), Undaria (2,359 thousand tonnes), Porphyra spp. (1,806 thousand tonnes), Sargassum (175 thousand tonnes) Spirulina spp. (86 thousand tonnes) and other aquatic plants (482 thousand tonnes).

PESTICIDES AND FERTILISER

The initial cultivation of crops began between 3,000 and 6,000 BC however, even before that, grain crops, in particular, were being grown as food crops. In the Old Testament, the Israelites, during the exodus, unfavourably compared their somewhat monotonous diet of manna with that from their time in Egypt when fish, cucumbers and leeks (and other alliums) were part of the diet (Numbers 11:5) While the initial cultivation of crops was certainly part of the agricultural revolution, the most significant expansion in crop and livestock yields came about between 1700 and 1900. The third agricultural revolution occurred in the 20th century and was a result of the use of fertilisers and crop protection agents as well as livestock growth promoters and veterinary medicine. We are now entering the fourth age of farming and cultivation. The growth of anti-microbial resistance due, in part to the prophylactic use of antibiotics in livestock and a range of issues associated with the use of pesticides and high phosphate and nitrate fertilisers means that a range of new approaches are being developed. These are designed not only to increase production yield (increased productivity) but also to work with the environment as a whole and to maintain sustainability and minimise environmental impact.

Advances in crop and livestock protection and fertilisers during the 20th century led to the green revolution and a substantial improvement in the resilience of the agri-food chain. The green revolution was fired by improvements in agronomic practice and in the development of new and better varieties. This resulted in a doubling of crop yields over a 50 year period in the 20th century and to the development of a number of institutions including the Consultative Group on International Agricultural Research (CGIAR) who were at the forefront of ensuring the dissemination of improved varieties [44]. More than 8,000 new varieties were produced in 11 target crops (rice, wheat, maize, sorghum, pearl millet, barley, beans, lentils, groundnuts, cassava and potatoes) over a 30 year period from the mid 1960s. It has been estimated that research related to the green revolution saved between 18 and 27 million hectares from being brought into agricultural use [45]. Since changes in land use is the third most important cause of carbon emissions globally, prevention of deforestation is an important aspect of the green revolution. By improving productivity on existing farm land, deforestation and loss of habitat can be reduced. However, it should be noted that increasing yields, particularly of valuable crops, may have the reverse effect and may encourage the conversion of forest, which is relatively unprofitable, into farm land, which provides an annual yield.

Apart from changes in crop varieties, prevention of losses due to insect predation and other pests including weeds and disease are an important aspect of increasing crop yield on the same acreage of cultivation. Pesticide usage on commercial crops began in Europe and North America in the late 1800s and early 1900s. While botanical preparations were used, many of these early pesticides were generic and contained heavy metals (arsenic, lead) or sulfur and petroleum. Accompanying them were inorganic agents such as sodium chlorate. The development of synthetic pesticides was a significant step in the control of pests on crops, however it should be remembered that a range of insecticides were, and still are used to protect animals and livestock from lice, ticks, fleas, intestinal parasites and nematodes. The most significant breakthrough chemical was dichlorodiphenyl trichloroethane – commonly known by its abbreviation, DDT. It was first synthesised by Ziedler in 1874 but was not known to be an insecticide until 1939 when this aspect of its biological activity was discovered by Paul Hermann Müller. While DDT is now banned in many countries, it should be remembered that Müller received the Nobel Prize for Physiology or Medicine in 1948 "for his discovery of the high efficiency of DDT as a contact poison against several arthropods" and was praised by Fischer in his Presentation Speech for the discovery that "preventative medicine is now able to fight many diseases carried by insects in a way that is totally different". Unfortunately one of the key benefits of many synthetic pesticides – their persistence, has also resulted in a major drawback. Accumulation of pesticides in non-target species with unanticipated effects as you move up the food chain has led to them being subject to regulation and, in many cases, their use has been prohibited. This was in a large part precipitated by Rachel Carson and her book 'Silent Spring'. Contrary to popular belief, Carson did not advocate a total ban on pesticides (including

DDT). As she notes “No responsible person contends that insect-borne disease should be ignored”. Her proposition was rather that uncontrolled use of pesticides had the dual effect of harming non-target species and allowing the target species to build up resistance. It is, essentially, an early example of what would be later manifested as anti-microbial resistance (AMR) which is now seen as a major global challenge. She would later comment “If Darwin were alive today the insect world would delight and astound him with its impressive verification of the survival of the fittest. Under the stress of intensive chemical spraying the weaker members of the insect population are being weeded out. Only the strong and fit remain to defy our efforts to control them”. While this is somewhat hyperbolic and is not the way resistance actually works – in the case of mosquitos, a single genetic mutation confers resistance to DDT and to pyrethroids – it reflects a common view and the end result is the same [46].

Attempts to mitigate the effects of pesticides have focused on improving specificity and reducing half-life and thereby lowering environmental impact. Unfortunately estimating half-life is difficult since it varies according to environmental conditions however some representative values for pesticide dissipation from plants are given in the Table 1 [47].

Table 1. Pesticide dissipation from plants

Pesticide	Half-life (days)	Date of Introduction
DDT	10.78	1944
Glyphosate (organophosphate)	3.98	1974
Chlorpyrifos (organophosphate)	4.01	1965
Malathion (organophosphate)	2.48	1956
Quintozene (fungicide)	14.74	1969
Thiamethoxam (neonicotinoid)	3.97	2002
2,4-Dichlorophenoxyacetic acid	2.92	1940

Given the spread of resistance to chemical agents and environmental concerns, there have been a number of attempts to develop alternative routes to new pest control agents. There have been significant efforts made in China and India to develop new molecules while others have looked at new formulations and the regulatory situation. It is, however, increasingly difficult to find new active chemicals. It has been estimated that almost 25,000 new compounds are screened for each new pesticide brought to market. Even 10 years ago, that figure was closer to 10,000. Many pesticides currently in use were developed over 30 years ago however major agrichemical multinationals typically spend 10 - 15% of sales revenue on R&D so there is still considerable investment in developing new approaches.

Apart from the use of genetic modification to aid pest control (which is subject to its own drawbacks), there have been a number of new approaches. The use of pheromones has been adopted in some environments however they tend only to be effective in specific environments and at low levels of infestation. They can also be used in conjunction with traps to determine the level of infestation. While pheromones have been researched for over 40 years as a pest control method, their use is still not widespread in major crops [48]. An alternative approach which has received some recent attention is the use of insect neuropeptides. Neuropeptides are involved in almost all physiological processes in insects including diuresis, ecdysis and pheromone biosynthesis. The concept is not new, having been proposed in the 1980s [49]. However as understanding of the role of insect neuropeptides has grown, the numbers of potential targets has increased. Virtually every aspect of an insect's life and life cycle is regulated by neuropeptides and there is sufficient variety between different insect species for the design of peptidomimetics to disrupt insect function to be specifically targeted to one species. Since neuropeptides are so essential and their receptors so important to insects, the development of resistance is unlikely. In addition, small peptide mimics are likely to be broken down rapidly in the environment. Finally, as is the case with pheromones, neuropeptides are extremely biologically active and only very small amounts will need to be applied to be effective.

In the case of rational design for herbicides, the situation is less clear. Many of the potential advances have (and will) involve engineering resistance to bioactive compounds into crop plants. The use of mutagenized plants and the identification of herbicide tolerant variants has been carried out [50]. This is only made possible by having a clear understanding of the mechanism of action of herbicides and may be made more efficient in future by using rapid screening of clones and novel crosses of wild varieties. This is an additional reason why preservation of non-commercial varieties may be so important.

The use of fertilisers to enhance crop growth and yields is an essential part of modern farming. There remain issues both with the availability of (particularly) potassium based fertilisers and run off leading to pollution of waterways and production of algal blooms. Work in the 19th and early 20th centuries led to the industrialisation of nitrate and ammonia production and this greatly supported the rapid growth in crop yields. Chemistry was at the core of these advances with the Haber and Ostwald processes being among the most significant. Further work will be necessary to optimise fertiliser application and use by crops and to prevent the consequences of run off and the depletion of natural resources.

Agricultural Policy

The key determinants of agricultural policy are the maintenance of a stable food supply and sustainability. In the EU the Common Agricultural Policy (CAP) is designed to create a stable and sustainable environment for farmers. In addition, there is a strong economic case for the policy since 44 million jobs in the direct food chain (food processing, retail and services) are dependent on agriculture. The EU is a net exporter of food and this is worth more than €13bn per annum. In contrast, the UK is a net importer of food and drink (-£22bn trade balance), like Norway and unlike Canada (CDN$ 11.5bn trade balance). The majority of UK food and drink exports go to EU countries (particularly Ireland, France, the Netherlands and Germany) so the effect of Brexit on this important facet of the UK economy will be likely to drive UK agricultural policy in the future.

For the EU, the CAP provides income support for farmers based upon the importance of the market and this is also linked to environmental sustainability, animal health and welfare and food safety. The CAP also seeks to minimise the impacts of the effects of weather conditions and market volatility on vulnerable European agricultural markets. The third key aspect of the CAP is that it seeks to respond to the needs for rural development in each of the member states. The future evolution of the CAP will prioritise climate change and preserving the environment. This will mean that the introduction of modern technologies, addressing consumer concerns, encouraging younger people to take up farming and ensuring the CAP is aligned with other EU policies on trade, migration and sustainable development. In addition, the EU is proposing to set up an EU-wide platform on risk management to enable farmers and the farming system to be more resilient. The fact that there will need to be changes to the CAP system is driven by changes in global conditions but also by the current economics of the system. Through the CAP, EU farmers collectively receive subsidies of about €52bn per annum. Of this, the majority (€40bn) is given in direct aid to farmers. It is unclear what will happen once the UK leaves the EU however since about 55% of total UK income from farming comes from CAP support including €3.1bn in direct payments to farmers, it seems likely that, even with the likely introduction of UK national subsidies, farm incomes will be reduced. In addition, if the UK continues to trade with the EU under a 'Norway-type' agreement, British farmers will still have to accept EU policies in areas such as the environment and this may disadvantage UK producers.

The US agricultural policy follows a five year cycle and deals with farming, food and nutrition, rural communities and bioenergy and forestry. The most recent was the 2014 farm bill which will remain in force through 2018. It is projected by the Congressional Budget Office that 80% of the spend will be in the area of nutrition and the total outlay will be $489bn. One key aspect of the support given to farmers is the crop commodity program. This provides benefits based on price or revenue targets for producers of corn or other feed grains, wheat, rice, soya beans, oilseeds, peanuts and pulses. There is also support for cotton producers. The US is a member of the OECD and provides information that allows the OECD to compare government support to agriculture across member (and selected non-member) countries.

The OECD itself provides information on policy developments and related support via an annual report. The report covers almost 90% of the global value added in agriculture. The latest version of the report notes that, despite shared goals across member states, there remains a lack of alignment of support for agriculture with the underlying objectives of government interventions. There are also significant market distortions. The OECD recommends redirecting producer support to more general agricultural service support thereby moving the sector away from production and trade distorting forms of support. The OECD bases its recommendations on an annual Producer Support Estimate (PSE) which tracks the monetary value of transfers to agriculture from consumers and taxpayers. The percentage total support estimate indicator represents the total of policy transfers to the agriculture sector as a % of GDP. The total support estimate for the EU in 2016 was 0.68% of GDP and the PSE was 21% of gross farm receipts. The US had a total support estimate of 0.49% of GDP and a PSE of 8.7%. The comparable figures for China were 2.2% of GDP and a PSE of 14.5% which may in part reflect the level of technological innovation in China.

EMPLOYMENT, HEALTH AND SAFETY

Approximately one third of the world's population are employed in the global agricultural sector, but this ranges from around 2% in the US to 80% in some African countries. Growth in the agriculture sector up to four times more effective in raising incomes for vulnerable individuals compared to other sectors, but in 2007, employment in the service sector overtook that of agriculture [51]. Although there are opportunities post-farm production (i.e., food processing, transportation, retailing, restaurants, and other services), these so called "pull" or "push" factors are leading to movement of people out of farming, leading to eventual skills deficits.

Apart from ensuring the safety of working conditions, safety in primary production is largely concerned with the safety of the produce itself. Nonetheless, there is a persistent number of accidents which can cause injury. According to the Health and Safety Executive, in the UK there were

27 fatal injuries to workers in the agriculture, food and fisheries sector in the UK in 2016/17. In addition, there were 15,000 cases of workers suffering from a work-related illness and 13,000 non-fatal injuries to workers during the same period. Both fatal and non-fatal injuries were caused by falls, injuries caused by animals and contact with machinery. Because fewer people work in agriculture than in many other industries (e.g., construction), agriculture had the highest rate of fatal accidents (8.6 fatalities per 100,000 workers compared to 1.8 fatalities per 100,000 in construction). In addition, the level of accidents in farming is not decreasing and research suggests that farmers' attitudes to safety only change after serious injury has occurred. The situation is even worse in the US where the fatality rate in 2015 was 19.2 deaths per 100,000 workers. In the EU as a whole the rate of fatal injuries in the sector was 9.4 per 100,000 workers in 2010 compared to 18.4 in the US in the same year. In addition to accidents, there is also an additional burden of occupational injuries caused by exposure to pesticides, muscoskeletal disorders and deafness caused by machinery.

Although there is a raft of regulations concerning the safe use of pesticides in an agricultural workplace, occupational exposure does occur. A recent study suggested that data from human biomonitoring as a tool to investigate exposure is lacking (Bevan et al., 2017). A range of measure will need to be adopted if human biomonitoring is to be implemented as part of the occupational health surveillance for pesticides in Europe. In terms of the effects of exposure on health, a review in 205 involving a small number (35) of agricultural workers found a positive association between occupational exposure to pesticides and biomarkers of lung health [53]. A number of studies have linked long term, chronic exposure to pesticides (organophosphates) with later cognitive impairment. The data was summarised in a systematic review carried out in 2013 [54].

A number of models have been developed to estimate exposure to pesticides both for those working with them and those whose contact is inadvertent. The models have been found to be sufficiently conservative. In addition, in April 2017, EFSA produced a report showing that over 97% of food samples analysed across Europe in 2015 were within the legal limits with 53% being pesticide free. In the case of animal products, 85% were pesticide free [55].

CONCLUSION

Malnutrition and food-borne diseases impose large current and future human, economic, fiscal and social costs on countries. Factors associated with poor diet (high systolic blood pressure, high body mass index, high fasting plasma glucose, diet high in sodium, diet low in fruits, high total cholesterol) are among the top risk factors for deaths world-wide with undernutrition remaining the leading risk factor for children under five [56]. Designing an agricultural systems to deliver improved nutrition and health requires a production-to-consumption approach. Improving agricultural practice to be more efficient is essential if we are to improve food security. Agriculture can boost the economy, reduce poverty and deliver employment and commercial opportunities, as well as empowering vulnerable communities. Underpinning this is the importance of protecting our natural resources and maintaining biodiversity. For agriculture to be effective in meeting the needs of the increasing population, it needs to move towards a more dynamic model. There should be more emphasis on increasing yield through sustainable practice and reduction of food loss and waste across all sectors, as well as considered trade and agricultural policy. Food produced should be fit for purpose, delivering key nutrients with minimised inputs in terms of land, water and energy. New opportunities in crop production and breeding, particularly non-cereal high protein plants are likely to help meet predicted nutrient gaps. In combination with improving soil health low-carbon, high-yield, resilient crops are a possibility. As outlined in the World Food Bank Strategy (REF) we need a "food system that is more resilient and that shifts from being a major contributor to climate change to being part of the solution". It is essential that our current agricultural skills and not lost, but supported by new skills and technologies. This will allow the development of digital and mobile platforms, facilitating innovative procurement methods. Chemical technologies will play an important role in achieving this objectives and this should be a priority for future research. It is clear that improvements in agriculture demands not just an understanding of livestock and crop production, but also geographies, climate adaptation, human nutrition, culture and complex supply chains, with chemical technologies contributing across all of these actors.

REFERENCES

[1] *The World Bank Health, Nutrition and Population dashboard.* http://datatopics.worldbank.org/health/population Accessed 2017.

[2] *OECD National Accounts Gross Domestic Product (GDP) and other OECD annual national accounts statistics.* http://www.oecd.org/std/na/gross-domestic-product-gdp-and-other-annual-national-accounts-statistics-oecd.htm Accessed 2017.

[3] *Global Employment Trends 2014: Risk of a jobless recovery?*/ International Labour Office. Geneva: ILO, 2014.

[4] GBD 2015 Obesity Collaborators Health Effects of Overweight and Obesity in 195 Countries over 25 Years *N Engl J Med.*, 2017, 377(1), 13-27. doi: 10.1056/NEJMoa1614362.

[5] FAO, IFAD and WFP. 2015. The State of Food Insecurity in the World 2015. *Meeting the 2015 international hunger targets: taking stock of uneven progress. Rome, FAO.*

[6] Hempson, GP; Archibald, S; Bond, WJ. The consequences of replacing wildlife with livestock in Africa. *Sci Rep.*, 2017, 7, 17196.

[7] Dallimer, M; Tinch, D; Acs, S; Hanley, N; Southall, HR; Gaston, KJ. and Armsworth PR 100 years of change: examining agricultural trends, habitat change and stakeholder perceptions through the 20th century *Journal of Applied Ecology*, 2009, 46, 334–343 doi: 10.1111/j.1365-2664.2009.01619.x.

[8] Max, Troell; Arne, Eide; John, Isaksen; Øystein, Hermansen. *Anne-Sophie Crépin Seafood from a changing Arctic Ambio.*, 2017 Dec, 46(Suppl 3), 368–386 UNFCCC. Adoption of the Paris Agreement. (2015).

[9] Bronson, W. Griscom; Justin, Adams; Peter, W. Ellis; Richard, A. Houghton; Guy, Lomax; Daniela, A. Miteva; William, H. Schlesinger; David, Shoch; Juha, V. Siikamäki; Pete, Smith; Peter, Woodbury; Chris, Zganjar; Allen, Blackman; João, Campari; Richard, T. Conant; Christopher, Delgado; Patricia, Elias; Trisha, Gopalakrishna; Marisa, R. Hamsik; Mario, Herrero; Joseph, Kiesecker; Emily, Landis; Lars, Laestadius; Sara, M. Leavitt; Susan, Minnemeyer; Stephen, Polasky; Peter, Potapov; Francis, E. Putz; Jonathan, Sanderman; Marcel, Silvius; Eva, Wollenberg; Joseph, Fargione. *Natural climate solutions Proc Natl Acad Sci U S A.*, 2017, Oct 31, 114(44), 11645–11650.

[10] Wei-feng, Zhang; Zheng-xia, Dou; Pan, He; Xiao-Tang, Ju; David, Powlson; Dave, Chadwick; David, Norse; Yue-Lai, Lu; Ying, Zhang; Liang, Wu; Xin-Ping, Chen; Kenneth, G. Cassman; Fu-Suo, Zhanga, *New technologies reduce greenhouse gas emissions from nitrogenous fertilizer in China Proc Natl Acad Sci U S A.*, 2013, May 21, 110(21), 8375–8380.

[11] Zou, J; Huang, Y; Jiang, J; Zheng, X; Sass, R. A 3-year field measurementof methane and nitrous oxide emissions from rice paddies in China: effects of water regime, crop residue, and fertilizer application. *Glob. Biogeochem. Cycles*, 2005, 19, GB2021.

[12] Zhisheng, Yao; Xunhua, Zheng; Chunyan, Liu; Shan, Lin; Qiang, Zuo; Klaus, Butterbach-Bahl. Improving rice production sustainability by reducing water demand and greenhouse gas emissions with biodegradable films. *Sci Rep.*, 2017, 7, 39855.

[13] Rosegrant, MW; et al. Looking into the future for agriculture and AKST (Agricultural Knowledge Science and Technology). In *Agriculture at a crossroads* (eds McIntyre B. D., Herren H. R., Wakhungu J., Watson R. T., editors.), 2009, pp. 307–376 Washington, DC: Island Press.

[14] World Bank. *Minding the stock: bringing public policy to bear on livestock sector development*. Washington, DC, 2009, Report no. 44010-GLB. [Ref list].

[15] Delgado, C. Rising demand for meat and milk in developing countries: implications for grasslands-based livestock production. In *Grassland:*

a global resource (ed. McGilloway D. A., editor.), 2005, pp. 29–39. The Netherlands: Wageningen Academic Publishers.

[16] Dumas, A; Dijkstra, J; France, J. Mathematical modelling in animal nutrition: a centenary review. *J. Agric. Sci.*, 2008, 146, 123–142.

[17] Borrelli, L; Coretti, L; Dipineto, L; Bovera, F; Menna, F; Chiariotti, L; Nizza, A; Lembo, F; Fioretti, A. Insect-based diet, a promising nutritional source, modulates gut microbiota composition and SCFAs production in laying hens. *Sci Rep.*, 2017, 7, 16269.

[18] Seungjib, Jeon; Jong-Min, Lim; Hyung-Gwan, Lee; Sung-Eun, Shin; Nam, Kyu Kang; Youn-Il, Park; Hee-Mock, Oh; Won-Joong, Jeong; Byeong-ryool, Jeong; Yong, Keun Chang. *Current status and perspectives of genome editing technology for microalgae Biotechnol Biofuels.*, 2017, 10, 267.

[19] Raun, AP; Preston, RL. History of hormonal modifier use. *Symposium: Impact of implants on performance and carcass value of beef cattle, P-957, Oklahoma Agricultural Experimental Station, Division of Agricultural Sciences and Natural Resources. Stillwater, OK: Oklahoma State University, 1997.*

[20] Biswas, S; Shapiro, CA; Kranz, WL; Mader, TL; Shelton, DP; Snow, DD; Bartelt-Hunt, SL; Tarkalson, DD; van Donk, SJ; Zhang, TCS. Ensley Current knowledge on the environmental fate, potential impact, and management of growth-promoting steroids used in the US beef cattle industry. *Journal of Soil and Water Conservation*, 2013, 68(4), 325-336.

[21] Philip, K. Thornton Livestock production: recent trends, future prospects. *Philos Trans R Soc Lond B Biol Sci.*, 2010, Sep 27, 365(1554), 2853–2867.

[22] Smith, C. Rates of genetic change in farm livestock. *Res. Dev. Agric.*, 1984, 1, 79–85.

[23] Elsik, CG; Tellam, RL; Worley, KC. *The Bovine Genome Sequencing and Analysis Consortium Science The Genome Sequence of Taurine Cattle: A Window to Ruminant Biology and Evolution*, 24, Apr 2009, Vol. 324, Issue 5926, pp. 522-528.

[24] *FAO. The State of World Fisheries and Aquaculture 2016.* Contributing to food security and nutrition for all. Rome, 2016, 200 pp.

[25] Hasan, MR; New, MB; eds. On-farm feeding and feed management in aquaculture. *FAO Fisheries and Aquaculture Technical Paper No. 583. Rome, FAO*, 2013, 67 pp.

[26] Salyer, SJ; Silver, R; Simone, K; Behravesh, CB. Prioritizing Zoonoses for Global Health Capacity Building—Themes from One Health Zoonotic Disease Workshops in 7 Countries, 2014–2016. *Emerg Infect Dis.*, 2017 Dec, 23(Suppl 1), S55–S64.

[27] O'Neill. The Review on Antimicrobial Resistance. *Tackling drug-resistant infections globally: final report and recommendations*, May 2016. https://amr-review.org/sites/default/files/160525_Final%

[28] Joint Committee on the Use of Antibiotics in Animal Husbandry and Veterinary Medicine (the Swann Report), *Cmnd.*, 4190, November 1969. London: Her Majesty's Stationary Office.

[29] Kahn, LH. *Antimicrobial resistance: a One Health perspective Transactions of The Royal Society of Tropical Medicine and Hygiene*, Volume 111, Issue 6, 1 June, 2017, Pages 255–260.

[30] Axel, Mie; Helle, Raun; Andersen, Stefan Gunnarsson; Johannes, Kahl; Emmanuelle, Kesse-Guyot; Ewa, Rembiałkowska; Gianluca, Quaglio. *Philippe Grandjean Human health implications of organic food and organic agriculture: a comprehensive review Environ Health.*, 2017, 16, 111.

[31] David, Makowski; Thomas, Nesme; François, Papy; Thierry, Doré. Global agronomy, a new field of research. A review *Agronomy for Sustainable Development*, April 2014, Volume 34, Issue 2, pp. 293–307.

[32] Godfray, HCJ; Beddington, JR; Crute, JI; Haddad, L; Lawrence, D; Muir, JF; Pretty, J; Robinson, S; Thomas, S; Toulmin, C. Food security: the challenge of feeding 9 billion people. *Science*, 2010, 327, 812–818.

[33] Tscharntke, T; Clough, Y; Wanger, TC; Jackson, L; Motzke, I; Perfecto, I; Vandermeer, J; Whitbread, A. Global food security,

biodiversity conservation and the future of agricultural intensification. *Biol Conserv*, (2012), 151, 53–59. doi: 10.1016/j.biocon. 2012.01.068.

[34] *EU (European Commission). Our Life Insurance, our Natural Capital: An EU Biodiversity Strategy to 2010. Brussels*, 3, May 2011.

[35] Lodge, DM; Williams, S; MacIsaac, HJ; Hayes, KR; Leung, B; et al. Biological invasions: Recommendations for us policy and management. *Ecological Applications*, (2006), 16, 2035–54.

[36] DAISIE. *Handbook of Alien Species in Europe, volume 3 of Invasion Ecology. Berlin (DE)*: Springer, (2009).

[37] Dupin, M; Reynaud, P; Jarošík, V; Baker, R; Brunel, S; Eyre, D; Pergl, J; Makowski, D. Effects of training dataset characteristics on the performance of models for predicting the distribution of Diabrotica virgifera virgifera. *PloS One*, (2011), 6, 1–11.

[38] Mueller, ND; Gerber, JS; Johnston, M; Ray, DK; Ramankutty, N; Foley, JA. Closing yield-gaps through nutrient and water management. *Nature*, (2012), 490, 254–257. doi: 10.1038/nature 11420.

[39] Lobell, D; Burke, M. On the use of statistical models to predict crop yield responses to climate change. *Agr Forest Meteorol*, (2010), 150, 1443–1452. doi: 10.1016/j.agrformet.2010.07.008.

[40] Philibert, A; Loyce, C; Makowski, D. Quantifying uncertainties in N2O emission due to N fertilizer application in cultivated areas. *PlosOne*, (2013), 7. doi: 10.1371/journal.pone.0050950.

[41] Ingram, JSI; Porter, JR. Plant science and the food security agenda, *Nature Plants*, (2015).

[42] Rasmussen, PE; Goulding, KWT; Brown, JR; Grace, PR; Janzen, HH; Korschens, M. Agroecosystem—long-term agroecosystem experiments: assessing agricultural sustainability and global change. *Science*, 1998, 282, 893–896.

[43] Eigenbrod, C; Gruda, N. Urban vegetable for food security in cities. A review. *Agron. Sustain. Dev.*, 2014, 35, 483–498.

[44] Pingali, PL. *Green Revolution: Impacts, limits, and the path ahead PNAS*, (2012), 109 (31), 12302-12308.

[45] Stevenson, JR; Villoria, N; Byerlee, D; Kelley, T; Maredia, M. *Green revolution research saved an estimated 18 to 27 million hectares from being brought into agricultural production PNAS*, (2013), 110, 8363.

[46] Riveron, Jm; Yunta, C; Ibrahim, SS; Djouaka, R; Irving, H; Menze, BD; Ismail, HM; Hemingway, J; Ranson, H; Albert, A; Wondji, CS. *A single mutation in the GSTe2 gene allows tracking of metabolically based insecticide resistance in a major malaria vector Genome Biology*, (2014), 15, R27, 2014.

[47] Fantke, P; Gillespie, BW; Juraske, R; Jolliet, O. Estimating half-lives for pesticide dissipation from plants. *Environ. Sci. Tech.*, 2014, 48, 8588–8602.

[48] *Methods in Biotechnology, Volume 5*, Ed F. R. Hall and J. R. Menn. Chapter 21, Pheremones for Insect Control. D R Thomson, L J Gut and J W Jenkins.

[49] O'Shea, M. Neuropeptides in insects: Possible leads to new control methods. In Von Keyserlingk, H, Jager, A, von Szczepanski C. Approaches to New Leads for Pesticides. *Proceedings in Life Sciences*, Springer, 1985.

[50] Shoba, D; Raveendran, M; Manonmani, S; Utharasu, S; Dhivyapriya, D; Subhasini, G; Ramchandar, S; Valarmathi, R; Grover, N; Krishnan, SG; Singh, AK; Jayaswal, P; Kale, Ramkumar PMK; Mithra, SVA; Mohapatra, T; Singh, K; Singh, NK; Sarla, N; Sheshshayee, MS; Kar, MK; Robin, S; Sharma, RP. Development and Genetic Characterization of A Novel Herbicide (Imazethapyr) Tolerant Mutant in Rice (Oryza sativa L.) *Rice*, 2017, 10, 10.

[51] Townsend, Robert. *Ending poverty and hunger by 2030: an agenda for the global food system*. Washington, D.C.: World Bank Group, 2015.

[52] Bevan, R; Brown, T; Matthies, F; Sams, C; Jones, K; Hanlon, J; La Vedrine, M. Human biomonitoring data collection from occupational exposure to pesticides. *EFSA External Scientific Report, 2017.*

[53] Mamane, A; Baldi, I; Tessier, JF; Raherison, C; Bouvier, G. Occupational exposure to pesticides and respiratory health. *European Respiratory review*, 2015, 24, 306-19.

[54] Ross, SM; McManus, IC; Harrison, V; Mason, O. Neurobehavioral problems following low-level exposure to organophosphate pesticides: a systematic and meta-analytic review. *Critical Reviews in Toxicology*, 2013, 43, 1, 21-44.

[55] European Food Safety Authority The 2015 EU report on pesticide residues in food, *2017 EFSA Journal*, 15(4), 4791134.

[56] Forouzanfar, MH; Alexander, L; Anderson, HR. (and others) Global, regional, and national comparative risk assessment of 79 behavioral, environmental and occupational, and metabolic risks or clusters of risks in 188 countries, 1990-2013: a systematic analysis for the Global Burden of Disease Study 2013. *The Lancet.*, 2015, Sep 11.

In: Food: The Vital Ingredient ISBN: 978-1-53614-402-4
Editors: A. B. Hanley and W. R. Russell

Chapter 3

PHYSICOCHEMICAL CHARACTERISTICS OF FOOD

Peter J. Wilde
Quadram Institute Biosciences, Norwich, England

INTRODUCTION

When considering nutrition and a healthy diet and what foods are healthy or unhealthy, most discussions naturally focus on the nutrients themselves since these are essential to a healthy life and their absence can be measured in terms of clinical outcomes. These nutrients include macronutrients: protein, carbohydrates and fat all of which are important for energy, building tissues and many other critical physiological functions. They also include micronutrients such as vitamins, minerals and other bioactive compounds in our diet that maintain health and prevent the onset of disease. However, little time is given to the fact that we don't consume nutrients in their pure form, we eat food. The food that we eat has a broad range of physico-chemical attributes that control how the food appears, how it tastes and feels in the mouth and most importantly how it digested, and how it releases the nutrients it contains. The physical chemistry of food relates more to how the individual chemical components of the food are assembled to form structures from the macromolecular scale upwards that give a food its unique characteristics These properties give foods their own distinctive tastes and textures, from the crunch of an apple or a biscuit, the softness of bread to the melting sensation of chocolate or ice cream. These physico-chemical attributes also allow foods to be transformed by cooking or processing from something that is inedible, into something that is tasty and sometimes nutritious. It was the inventiveness of our ancestors to transform foods into something more edible that initially solved our early societal problems of food shortage and distribution. However, this ability is now presenting us with the opposite problems of over consumption driving increased levels of obesity and metabolic disorders, whilst under-nutrition and food shortages are still affecting many developing countries. This chapter describes how the physico-chemical properties of food has changed through history and how this has impacted on the health and wellbeing of society, and how we can learn from this historical perspective to help tackle the global health issues that we are currently facing.

Evolution of Food Structure

From an evolutionary perspective, for over 99% of human history we have existed as hunter gatherers, eating whatever food was available including fruits, vegetables, seed, nuts and whatever fish and meat was available locally[1]. The foods consumed by our hunter-gatherer ancestors were often very difficult to digest. Our digestive system has therefore evolved over several million years to digest these foods and extract as many valuable nutrients in as large quantities as possible [2]. For example, nuts and seeds are designed to pass through the digestive system intact to allow plants to disperse and grow, this is due to the structure and composition of the cell walls in these foods that resist the action of digestive enzymes, and thus survive the digestive process [3]. The earliest forms of food processing grinding and cooking, transformed our diets and our development [4]. This made it possible to alter the chemistry and structure of the foods to neutralise toxins and pathogens and make some foods both more edible and more digestible and hence had a bigger impact on our nutrition and health. This enabled our early ancestors to thrive in a broad range of environments and survive on a broad range of diets. An essential aspect of a hunter-gatherer lifestyle was its seasonality, meaning that at certain times of the year, food was scarce. To allow for this, our digestive control systems developed to take on more food than we needed to so that the energy could be stored as fat to allow us to survive through periods when food was scarce [5]. These control mechanisms express themselves as hunger or fullness to control our eating behaviour. This drive to overconsume foods is still present in our genetic makeup and is thus an important factor that drives the development of obesity in modern humans [5].

The development of agriculture over 10-12,000 years ago together with food storage and distribution systems and eventually food preservation methods, allowed our population groups to grow and thrive, giving us a more assured and consistent food supply [1]. Milling and baking of cereal grains was an essential part of human development thought to have started around 30,000 years ago [6]. Whole cereal grains contain a lot of energy on the form of starch, but this is trapped in indigestible plant cell walls. So grinding or milling would help break up the grain, rupturing some of the cell walls and making the starch available. However, this raw starch can still be difficult to digest, so cooking the flour is required to gelatinise the starch, transforming the structure and physico-chemical properties that not only creating a much more edible food, but one that is also much more digestible and allows us to absorb much more of the available energy.

Further, as our population levels have spiralled upwards, particularly in dense, urban populations, the demands on food supply increased, and our food production and processing and distribution systems have had to become more and more efficient to meet that demand. This drove the growth of the food processing industry in the 19th and 20th centuries. Early processed foods were driven by military activities, to produce foods that could be stored, transported and prepared in order to supply armies with the essential nutrition required to keep military personnel fit and healthy. As the industry evolved, new food processing technologies were developed, such as extrusion cooking, producing a wider range of new foods and tastes. The use of salt, refined sugar and the discovery of monosodium glutamate in 1908 led to the production of tasty foods, where the aim was to provide consumers with energy and essential nutrients in a tasty and delicious form. This was particularly important after the 2nd world war, when food was often scarce, and food processing and preservation techniques were important to provide the population with sufficient food from limited raw materials. This in part led to the widespread use of dry ingredients, which could be stored for prolonged periods, then processed into foods to meet demand. These dry, refined ingredients had none of the structure or physico-chemical properties of the original raw materials, thus their breakdown during digestion will be completely different.

Physicochemical Properties of Food

The physical chemistry and structure of food and ingredients are essential aspects that must be considered to address global issues of food security, nutrition and health. As described above the form and structure of the food we eat has changed out of all proportion over the course of human history, and the efficient supply of safe, nutritious food has helped drive and sustain huge increases in global population. To maximise efficiency, raw materials and ingredients need to be in a form that can be stored for significant periods, will survive transportation, and can be transformed into the final food in local production facilities. Whole grains, such as wheat, rice and maize, together with legumes and seeds are the natural solution to this problem, and when the seed or grain matures, the water activity is kept low because of the physical chemistry of the main components. The starch, protein and lipids, are structured in a way to maintain integrity and viability, but minimise water activity. They can be stored and transported to ensure a consistent supply of raw materials for processing into foods.

The other approach is to separate and extract the raw materials into the key components of starch, protein and fats and other carbohydrates, that can be dried to extend storage stability. This allows a broader range of food products to be manufactured with different structures and textures. These foods are given desired textures through the use of functional ingredients such as emulsifiers, stabilisers and thickeners, from a wide variety of raw materials that allows different structures and textures to be created. However, this extensive processing of food and ingredients completely changes nutrient bioavailability and can have a demonstrable impact on health. As described earlier our ancestral diets comprised of raw, unprocessed foods, and much of the energy and nutrients were derived from plant based foods. The nutrients and energy from these foods are particularly difficult to extract, which is why humans have evolved a long gastrointestinal (GI) tract, including microbial fermentation in the colon to further breakdown the plant tissue structures that our endogenous enzymes cannot. The human GI tract has also developed complex neuroendocrine feedback and control mechanisms to control gut motility, appetite and secretions to optimise conditions to extract as much of the nutrients from the food as possible [7]. Therefore when the original structure of the food is destroyed through processing, all the nutrients become more digestible and rapidly available for absorption. This also has a knock-on effect, in that once the food has been digested, appetite returns more rapidly. So for highly bioavailable, processed foods, hunger will return sooner compared to a structured, less digestible food [8]. This can lead to overconsumption and weight gain and eventually to obesity and related metabolic disorders associated with chronic overconsumption. This is further complicated by the fact that it is thought that human digestive control systems have developed to allow moderate overconsumption as the norm [5]. This would allow our ancestors to store energy as fat when food was abundant, which would sustain the body during winter or times of famine. In modern, developed society, the existence of a complex food supply chain ensures that there are no periods low food availability, and so this moderate overconsumption continues throughout life leading to increased body weight as we age. The UK statistics on obesity clearly show this on Figure 1.

One example of how the structure of food can impact consumption and health risk is demonstrated by analysis of large prospective cohort studies. One such analysis looked at a very simple processing effect, comparing the consumption of whole fruit versus fruit juice [10]. They found that the consumption of whole fruits was associated with a reduced risk of developing type 2 diabetes. In contrast, consuming the same fruits in the form of a juice, led to an increased risk, due to the high bioavailability of free sugar. This has led the WHO to propose thatfruit juice contributes towards free sugar intake, but whole fruits do not [11]. Although the causal link is probably more complex than simply the availability of free sugar, this clearly demonstrates that a simple processing step can have a real impact on health outcomes.

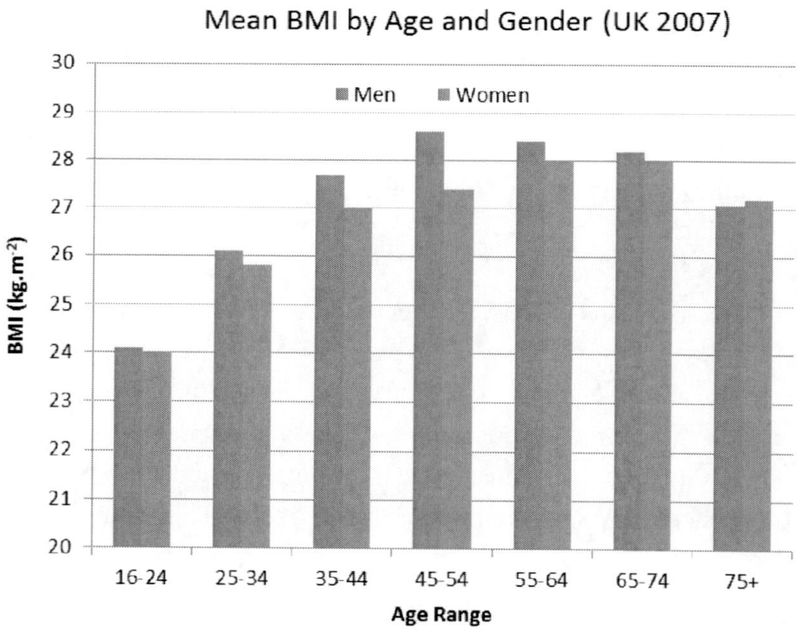

Figure 1. Mean BMI of men and women in the UK (2007) as a function of age. Data from UK Statistics on Obesity, Physical Activity and Diet: England February 2009 [9].

Although food processing can create a whole range of structures and textures to make food more pleasurable to eat, the digestibility and nutrient availability is something that needs to be considered to tackle the rising tide of obesity and type 2 diabetes. Another important issue is the rising global demand for protein, and one class of food whose structure is very difficult to replicate using raw ingredients, is meat.

The Global Meat Market

One major global food security issue, is the rising global demand for meat-based protein putting existing resources under tremendous strain. This has been referred to in other chapters. From a production standpoint, to produce 1kg of beef requires around 15,000 L of water, 6 kg grain and generates more CO_2 than production of any other meat or crop. The demand for meat is mainly driven by its taste and texture, consumption of meat far outweighs the nutritional requirements for protein, therefore there are moves to encourage consumption of other sources of protein, such as those from vegetable sources and the recent emerging utilisation of insects as a more sustainable source of protein. However these changes are happening very slowly, and the gains made in moving people away from meat is not reducing the global demand, which, according to FAO the per capita consumption is expected to double by 2030 from levels consumed in the 1960s, coupled with the rising global population, meat consumption continues to rise at a significant rate, and has increased by 60% since 1995 [12].

Altering consumer behaviour is very difficult, so persuading consumers to eat less protein, or to shift away from animal protein is extremely challenging. So the approach is to utilise processing methods to alter the physicochemical properties of alternative protein sources in order to give them the appearance and texture and organoleptic properties of meat. Hence there has been an growth in the market for vegetarian meat substitutes [13]. Global sales are predicted to rise from £2.9 billion in 2015 to £4 billion in 2020. Protein rich legumes such as soy and pea are regularly used to produce protein rich products of which tofu is probably the best known product. Tofu is thought to have originated in Northern China over 2000 years ago, is made from fermented soya bean milk in which the proteins are coagulated by a combination of heating and added coagulants. The heating denatures the proteins and makes them more susceptible to coagulation. The coagulants used will vary depending on the type of tofu made and the country of origin. Salts are traditionally used in Chinese tofu production, usually calcium or magnesium based, they screen neutralise the negative charges on the protein which normally prevent aggregation, thereby promoting coagulation. More modern methods also use other coagulants such as organic acids like glucono delta-lactone (GDL). These work by an alternative mechanism - they reduce the pH towards the iso-electric region of the main proteins, which negates the repulsive negative charge on the proteins and promotes coagulation. They are employed during the manufacture of cheese and yoghurt. Proteolytic enzymes such as papain are also used, to remove peptide chains that are sterically stabilising the proteins in solution. The coagulants are often used in different combinations by food manufacturers to create different textures and consistencies. However, few regard tofu as a meat substitute, it is a product in its own right, and often used to impart a particular texture in certain foods. More recently, food processing methods have been developed to transform vegetable proteins into structures that resemble the fibrous structure and texture of meat based products. So called Textured Vegetable Protein (TVP) is most commonly created using soy protein and was developed in the 1960's. Although more recently, other plant protein sources are also used including peas, wheat, cotton seed and oats. Most plant storage proteins are high molecular weight and globular and often formed from several subunits. To transform them into fibrous structures requires a combination of high temperature, pressure and controlled shear forces. Extrusion cooking is used to create the specific conditions that promote the denaturation and aggregation of proteins. The shear forces and dramatic pressure change that occurs during extrusion cooking aligns the protein aggregates into fibres, creating a high protein ingredient, that can be used as a base for processed foods.

Another method of creating meat-like fibrous structures is the use of myco-protein or fungal protein. The particular form of mycoprotein used in foods was developed in the 1960s to address concerns over population growth and global food shortages. Commercial myco-protein is formed from a filamentous fungi (Fusarium venenatum). The fungi is grown in vats to form a dense network of filaments and then processed using heating and freezing steps which creates bundles of fibres akin to those of meat.

The use of processing to create protein rich structures with physico-chemical properties similar to those of meat is an important approach to reduce our global reliance on meat protein, without having to significantly change consumer behaviour.

DELIVERING OPTIMAL NUTRITION

Malnutrition is a complex, global issue that affects many developing countries. In early life, undernutrition of essential nutrients can lead to stunting, failure to thrive and even death. In Africa, almost 60 million (over 25%) children under 5 are stunted [14], and deficient in micronutrients such as vitamin A, calcium and iron) [15]. Paradoxically, obesity and other lifestyle-related diseases are also prevalent in adulthood. The levels of obesity and type 2 diabetes are increasing in all African countries. This is giving rise to the situation where overweight and obese mothers are giving birth to stunted children [16]. This is partly because the major cereals (especially maize, wheat and rice) that constitute the main source of energy in most households in Africa, cannot meet the special nutritional requirements, notably folic acid, calcium, iron and zinc, of pregnant women and children. An increasingly urbanised population is changing its diet away from traditional, nutritionally rich crops, to foods from high yielding, nutritionally poor crops. This contributes to both undernutrition, particularly in infants and pregnant/lactating mothers, but also to a dramatic increase in the incidence of type 2 diabetes [17] (See Chapter 1 for a discussion of thrifty baby syndrome and early life metabolic programming). This has been termed as the double burden of malnutrition [18] that is affecting many countries throughout the developing world. This is clearly demonstrated in Figure 2, showing that the incidence of diabetes is now greater in low income countries than it is in high income countries.

The physico-chemical and structural properties of the food play an important role. Even in traditional foods that are rich in minerals and micronutrients, the bioavailability of these nutrients may be poor, as these nutrients are often locked up in the tough outer bran and aleurone layers [19], and therefore not bioavailable. Also the presence of antinutrients such as phytates in the whole grains can inhibit the uptake of minerals [20]. At the other end of the spectrum, energy dense crops such as maize and wheat comprise over 60% starch, which is trapped within endosperm cells. However, endosperm cells have much weaker cell walls, and milling is effective at breaking down the endosperm structure. Therefore the starch in foods made from milled flour is highly bioavailable. This consumption of energy dense foods is driving the increasing incidence of obesity and type 2 diabetes being observed in many developing countries [17].

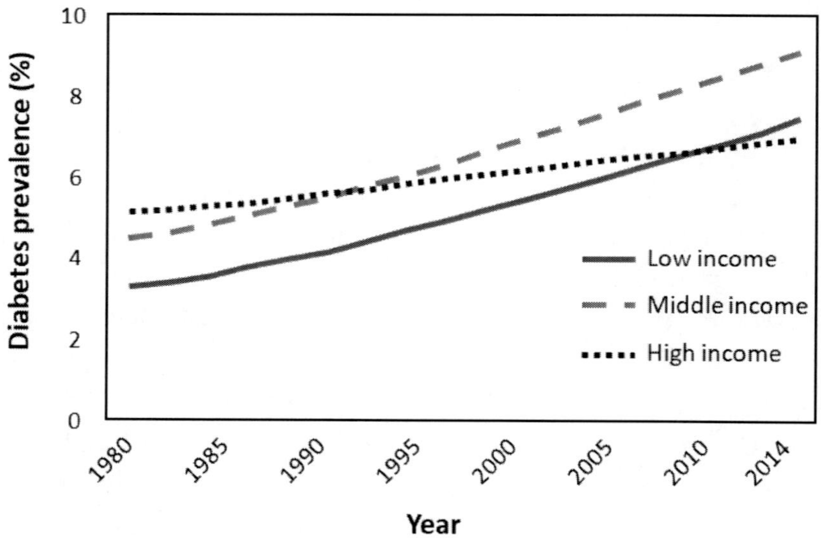

Figure 2. Diabetes prevalence by year in low, middle and high income countries (data from WHO Global report on diabetes [17]).

The structure and physical chemistry of the food could therefore play an important role in improving the nutritional profile of foods to suit different life stages. As discussed above, we need to address problems of undernutrition in early life, yet also combat the consequences of lifelong over consumption in adulthood. There are various approaches by which essential nutrients can be made more available. For example, fermentation processes can utile a range of fungal and bacterial enzymes to break down the cell wall polysaccharides such as pectins, arabinoxylans etc, breaking down the cell wall structure and liberating nutrients trapped inside cells. Fermentation has been shown to increase the bioaccessibility of minerals from cereals [21], and the process can be optimised to ensure the appropriate combinations of enzymes are expressed to breakdown the cell wall structures. Specialised milling approaches such as micromilling has been shown to increase the bioavailability of micronutrients [22] as it is a more effective way at breaking down the tough aleurone and bran layers in cereals that contain many essential micronutrients.

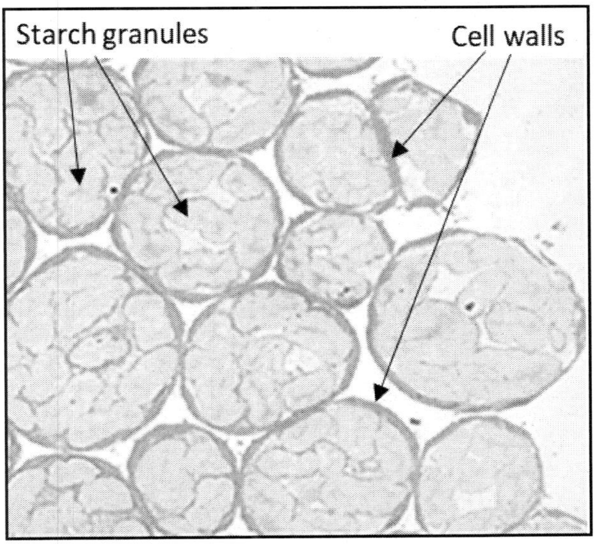

Figure 3. Cooked pea tissue that has passed through a simulated digestion process, showing intact cells containing gelatinised starch granules.

Similar approaches are also being investigated to obtain the opposite effect and modulate the release of energy from foods during digestion. For example, nuts and seeds have very tough cell walls that are highly resistant to both digestion and fermentation [3]. In fact, many seeds have evolved to be dispersed by passing through the digestive system of animals. The energy content of many seeds and nuts is high, however, and they are linked with inducing satiety and promoting weight loss. This is thought to be due to their poor digestibility and slow energy release. Recent work on almonds showed that only 60% of the energy is actually absorbed [23], and many intact cells, along with their contents, can be observed in human faecal samples [3]. Thus the cell walls are important structures controlling energy release. Figure 3 shows pea tissue that has gone through an in vitro digestion, showing intact cells containing starch. The slow rate of digestion means that undigested material will persist to the end of the small intestine, and their slow release of nutrients are sensed by the GI tract, signalling to reduce gut motility and enhance satiety. The rate of digestion can be simply controlled by altering the particle size of the material, such that a food that has been milled into smaller particles has a greater surface area from which to release nutrients [24]. Therefore digestibility and nutrient release could be controlled by controlling the particle size of the raw materials.

On a smaller length-scale, the digestibility of starch can be controlled by the structure and composition of the starch granule itself. Starch is the single largest source of energy in our diet. According to the WHO over 50% of global energy intake is from cereal starch alone, yet it is free sugars that have been linked with obesity and related disorders. Therefore adjusting the levels and type of starch in our diet could potentially have a much bigger impact on our health. Starch is essentially a homopolymer of glucose molecules used as an efficient means of storing energy in plants. In humans it is digested by a-amylase secreted in saliva and by the pancreas and by gluco-amylase to break down the starch into maltose and maltotriose. These sugars are further broken down into glucose by enzymes in the brush border of the gut. The molecular and supra-molecular assembly of starch can strongly influence its digestibility by a-amylases. Raw native starch is often poorly digestible due to its low water content restricting access of amylases to the internal structure of the starch. Cooked starches are generally more digestible, as the starch has gelatinised and swollen, opening up the structure so that it can be hydrolysed by amylase. There are two main forms of starch, amylose and amylopectin. Amylose is a linear polymer of glucose linked through α(1-4) glycosidic bonds, with few branches which tends to form crystalline structures of close packed chains which are resistant to hydrolysis by amylase. Amylopectin is more branched (α(1-6) bonds), so when gelatinised, it forms a swollen, hydrated structure which allows amylase access to the core structure. Depending on the composition, structure and processing, the starch can either be rapidly degraded, giving rise to large peaks in blood glucose, more slowly degraded leading to a longer, sustained release of energy [25], or it can be resistant to digestion by endogenous enzymes. This resistant starch (RS) is then fermented by bacteria in the colon to produce short chain fatty acids that have a range of health benefits [26], as well as leading to the growth and diversification of the microbial population, which helps maintain a healthy gut and helps prevent a range of gastrointestinal disorders. The degree of resistance to digestion will therefore be dependent on the proportion of amylose, known as RS2. It can also depend on how the food has been processed. Some starches when cooled after cooking, can readily recrystallize (retrogradation). This retrograded starch is much higher in RS (RS3), for example, cooling cooked potatoes can significantly reduce starch digestibility [27] and increase the proportion of RS. Research is also ongoing to try and increase the amount of RS in the raw material, through advanced genomic technologies and breeding methods to increase the proportion of amylose in commonly consumed crops such as wheat [28] and rice [29].

CONCLUSION

There is great potential therefore to manipulate the physicochemical properties of food to improve the global nutritional impact and deliver appropriate nutrition to larger sectors of the population. These approaches can be used to address both undernutrition in terms of improving the bioavailability of essential nutrients as well as moderating the impact of energy dense foods on metabolic disorders such as type 2 diabetes. No single approach can successfully tackle these global issues alone and so requires a multidisciplinary approach from primary production (breeding and agronomy) through to processing and preparation methods. To have a significant influence on global health and food security, these approaches need to be incorporated into commonly consumed staple foods without significantly affecting their acceptability, which in itself can be a major barrier to the successful introduction of healthy foods into the market.

REFERENCES

[1] Krebs, J., *Food: A Very Short Introduction*. 2013, Oxford: Oxford University Press.

[2] Cordain, L., et al., Origins and evolution of the Western diet: health implications for the 21st century. *American Journal of Clinical Nutrition*, 2005. 81(2): p. 341-354.

[3] Ellis, P. R., et al., Role of cell walls in the bioaccessibility of lipids in almond seeds. *American Journal of Clinical Nutrition*, 2004. 80(3): p. 604-613.

[4] Zink, K. D. and Lieberman, D. E. Impact of meat and Lower Palaeolithic food processing techniques on chewing in humans. *Nature*, 2016. 531(7595): p. 500-+.

[5] Wells, J. C. K., The evolution of human fatness and susceptibility to obesity: an ethological approach. *Biological Reviews*, 2006. 81(2): p. 183-205.

[6] Henry, A. G., Brooks, A. S. and Piperno, D. R. Microfossils in calculus demonstrate consumption of plants and cooked foods in Neanderthal diets (Shanidar III, Iraq; Spy I and II, Belgium). *Proceedings of the National Academy of Sciences*, 2011. 108(2): p. 486-491.

[7] Drucker, D. J., The role of gut hormones in glucose homeostasis. *Journal of Clinical Investigation*, 2007. 117(1): p. 24-32.

[8] Venn, B. J. and Mann, J. I. Cereal grains, legumes and diabetes. *European Journal of Clinical Nutrition*, 2004. 58(11): p. 1443-1461.

[9] Statistics on Obesity, *Physical Activity and Diet: England*. 2009, The Information Centre.

[10] Muraki, I., et al., Fruit consumption and risk of type 2 diabetes: results from three prospective longitudinal cohort studies. *Bmj-British Medical Journal,* 2013. 347: p. 15.

[11] *WHO*. http://who.int/nutrition/publications/guidelines/sugar_intake_information_note_en.pdf. 2015.

[12] *FAO*. World Agriculture: Towards 2015/2030. An FAO perspective... 2003; Available from: http://www.fao.org/docrep/005/y4252e/y4252e00.htm#TopOfPage.

[13] Ridler, J. *Meat substitute market to grow to £4bn. 2017*; Available from: http://www.foodmanufacture.co.uk/Ingredients/Meat-substitute-market-to-reach-4.bn-globally.

[14] Bank, U. W. W. *Levels and Trends in Child Malnutrition: Key Findings of the 2015 Edition*. 2015; Available from: www.who.int/nutgrowthdb/jme_brochure2015.pdf?ua=1.

[15] Nutrition Country Profile – Kenya, in *Global Nutrition Report* 2015.

[16] Kimani-Murage, E. W., et al., Evidence of a Double Burden of Malnutrition in Urban Poor Settings in Nairobi, Kenya. *Plos One,* 2015. 10(6): p. 17.

[17] *WHO, Global report on diabetes*. 2016.

[18] *WHO.* Double burden of malnutrition. Available from: http://www.who.int/nutrition/double-burden-malnutrition/en/

[19] Baye, K., Guyot, J. P. and Mouquet-Rivier, C. The unresolved role of dietary fibers on mineral absorption. *Critical Reviews in Food Science and Nutrition*, 2017. 57(5): p. 949-957.

[20] Zhou, J. R. and Erdman, J. W., Phytic Acid in Health and Disease. *Critical Reviews in Food Science and Nutrition*, 1995. 35(6): p. 495-508.

[21] Obinna-Echem, P. C., Beal, J. and Kuri, V. Effect of processing method on the mineral content of Nigerian fermented maize infant complementary food - Akamu. *Lwt-Food Science and Technology*, 2015. 61(1): p. 145-151.

[22] Latunde-Dada, G. O., et al., Micromilling Enhances Iron Bioaccessibility from Wholegrain Wheat. *Journal of Agricultural and Food Chemistry,* 2014. 62(46): p. 11222-11227.

[23] Gebauer, S. K., et al., Food processing and structure impact the metabolizable energy of almonds. *Food & Function*, 2016. 7(10): p. 4231-4238.

[24] Grundy, M. M. L., et al., Impact of cell wall encapsulation of almonds on in vitro duodenal lipolysis. *Food Chemistry*, 2015. 185: p. 405-412.
[25] Lovegrove, A., et al., Role of polysaccharides in food, digestion, and health. *Critical Reviews in Food Science and Nutrition*, 2017. 57(2): p. 237-253.
[26] Lockyer, S. and Nugent, A. P. Health effects of resistant starch. *Nutrition Bulletin*, 2017. 42(1): p. 10-41.
[27] Tian, J. H., et al., Effects of preparation methods on potato microstructure and digestibility: An in vitro study. *Food Chemistry,* 2016. 211: p. 564-569.
[28] Hazard, B., et al., Induced Mutations in the Starch Branching Enzyme II (SBEII) Genes Increase Amylose and Resistant Starch Content in Durum Wheat. *Crop Science*, 2012. 52(4): p. 1754-1766.
[29] Sun, J., et al., Characterisation of starch during germination and seedling development of a rice mutant with a high content of resistant starch. *Journal of Cereal Science*, 2015. 62: p. 94-101.

In: Food: The Vital Ingredient ISBN: 978-1-53614-402-4
Editors: A. B. Hanley and W. R. Russell

Chapter 4

Challenges to Food Safety and Quality: Technical Challenges

Mike Saltmarsh
Inglehurst Foods Ltd., Hampshire, UK

Introduction

Food issues are dramatically different across the world, with the industrialised countries facing an obesity crisis while a considerable proportion of the population of other countries suffer malnourishment and famines. Since the Second World War the food industry in the developed world has been extremely successful in providing food which is safe, tasty, inexpensive and readily available. Unfortunately this low cost and the attractiveness of the food, combined with a decrease in the activity required for daily life, has contributed to the rise in obesity currently visible in Europe and America. The industry now has to find ways of providing food which is lower in energy content, yet is still tasty, contains all the nutrients necessary for daily life and is able to satisfy the hunger of consumers. This chapter will focus on the challenges facing the food manufacturing industry in the developed world.

The world food supply is facing two major challenges – population growth and climate change. The population of the world is growing and predicted to reach 9.7 billion by 2050 [1] so that there will be increasing demand for basic foodstuffs. That the climate is changing is accepted by the scientific community and the changes are becoming evident in terms of warming, rainfall and severity of extreme events [see for example Gardening in a changing climate RHS 2017, 2]. These issues will impact on the supply of foodstuffs into the food industry.

Against this difficult background the food industry in the developed world faces the challenges of meeting consumer needs; reducing cost, increasing convenience, new tastes and textures - but now in an environment where consumers are risk averse, poorly educated in science, heavily influenced by emotion on social media, encouraged to "eat natural" and to reject both additives and products from modern technological processes. It is worth noting that many improvements in food safety in the past have been opposed, with the retail trade against Victorian laws on adulteration, many in the dairy industry opposed to the introduction of pasteurisation of milk and the emotive arguments against irradiation as a sterilisation treatment for foodstuffs.

This chapter will discuss some of the technological issues that will need to be addressed throughout the food chain in our desire for a safe, high quality food supply in the future.

Basic Food Production

The issues surrounding increasing food production in a world of growing population and affected by global warming have been summarised by Krebs [3] and in the Lancet [4]. The most important macronutrients for all plants are water, nitrogen, potassium and phosphorus. Supplies of three of these; water, potassium and phosphorus are limited but could be enhanced by creative use of waste.

Because 70% of fresh water supplies in the world are exploited at the present time it is clear that with changing climate and, in particular, changes in rainfall patterns, the world is going to have to improve fresh water use and storage or find significant alternatives sources of fresh water by, for example, desalination. Agriculture is a major user of water and increasing use will need to be made of "grey" water for irrigation. However, grey water has by its nature been used once and may contain chemicals, bacteria or viruses that need to be removed before it can be safely used again. There have already been outbreaks of illness, including hepatitis A, attributed to the use of grey water for irrigating fruit crops. With the encouragement to eat more unprocessed food, particularly fruit and vegetables, the risk of illness increases and the challenges are to develop improved processes for cleaning water for reuse and for detecting contamination both in the water and in the crop. Many years of study have gone into providing clean mains drinking water to developed countries and the controls are well established. The use of indicator organisms in monitoring the quality and safety of mains water control is proven over 100 years but each type of grey water is unique and much needs to be done to establish safe treatment and monitoring systems for irrigation water which may be sub surface drip feed or overhead spray. This requires long term studies which the current funding systems are not well equipped to support. It is worth noting that waste water treatment processes have been developed to deal with historic contaminants but they are now facing challenges in dealing with low levels of pharmaceuticals, endocrine disruptors and nano materials, particularly nano silver, used in cosmetic and household use materials. The problems that can be caused by this last has already been demonstrated with the use of microplastic beads in body washes which have passed through the system and now contaminate the oceans. Meeting the challenges caused by increasing uses of new materials will be demanding for chemists, microbiologists and engineers.

The second most important nutrient for crops is nitrogen, and while it is readily available as an industrial chemical, it can provided by sewage sludge and the solid residue from some anaerobic digestion plants. These provide both nitrogen and humus for soil improvement but these residues need to be carefully managed in order to minimise the risks from harmful materials such as pathogenic bacteria or heavy metals. The practice of spreading this material on fields is well controlled at present but increasing amounts of sludge are being applied to the land and it is likely that we will need to develop techniques of more active removal of heavy metals in particular. As with the removal of arsenic from water used for irrigation of rice and other crops in Bangladesh, the challenge is not just to remove the metal but what to do with it once it has been removed.

Phosphorus is also present in sewage. At present the waste water treatment industry has to remove phosphate from water before it can be returned to rivers since phosphate can encourage the growth of algal blooms that can remove the oxygen from water with the subsequent death of invertebrates and fish. Phosphate is a mineral fertiliser that is a finite resource – there are few phosphate mines in the world. It is predicted that phosphate reserves could be exhausted in roughly 35 years unless new significant sources are discovered. The optimisation of the removal and concentration of phosphate from sewage so that it can be applied to fields will be an objective for the water treatment industry and will make some contribution to the provision of phosphate to farms [5].

Cereal crops such as wheat, rye and barley are susceptible to attack by fungi of both ergot and fusarium species which both reduce yields and produce harmful toxins. The limits on the amounts of these toxins present in foodstuffs are being reduced and while this may reduce the risks to consumers, unless agriculturalists are able to develop crops that are less susceptible to the fungi in the warmer, damper climate, this will merely result in more cereals being wasted. In addition, faster, more accurate methods for the determination of the fungi (and their toxic secondary metabolites) are needed. Currently higher levels are permitted in animal feed and studies are needed on where these toxins end up in the food chain. The industry also faces the challenge of what to do with crops that contain too high a level of contaminant. Unless there is a clear route for making good use of such crops, with the increasing demand for food, they are likely to enter the black economy.

A further challenge for growing crops is random environmental contamination. Dioxins arising from uncontrolled burning of plastics has given rise to contaminated pork in Europe and guar gum in India. It can be confidently predicted that while industrial combustion plants in Europe and

America will be improved to minimise the risks of airborne dioxins, small pockets of contamination will continue to arise in less developed countries. The development of portable equipment able to identify contamination is a challenge for analytical chemists and for the industry as a whole.

The interaction of human health with agricultural production is demonstrated most clearly in the impact of the use of antibiotics in animal husbandry. For some years the use of prophylactic antibiotics in the intensive production of chicken, ducks and pigs, has been routine to maximise rates of growth and conversion rates but it is now accepted that this has increased the speed at which bacteria have become resistant to the antibiotics upon which human infection treatment depends. Again, this practice can be controlled in the developed world, and regulation of the use of antibiotics as growth promoters phased out their use in Europe between 2002 and 2006 and the FDA Veterinary Feed Directive banned the use of antibiotics at sub-therapeutic doses as an aid to growth and improved feed efficiency from the 1st January 2017. However the use of antibiotics for improved growth efficiency is likely to continue in smallholder farms in Asia Pacific countries. Research is needed into means of maintaining the rates of growth and efficiencies without the use of antibiotics. Because this is a worldwide problem, the solution has to involve systems that are applicable to smaller farmers in Asia as well as efficient farms in Europe and the USA.

LOGISTICS

According to the FAO, approximately one third of all food produced is wasted [6]. Losses vary by crop, from 20% of oilseeds and dairy products to 45% of fruit and vegetables. They are spread across the food chain from production to consumer and almost all relate to perceived quality. While it is clear that better education of consumers on the meaning of expiry dates, whether for food safety or for quality, would help them to waste less food that is perfectly edible, wastage also occurs on the way from the farm to the store. Part of the solution is improved stock control and an acceptance that the current situation in supermarkets where all products have to be in stock all the time means that, for some short shelf life crops, excesses are produced and some either never leaves the field or is discarded. For much of the world, though, the solution is in better on-farm storage and the introduction of cold chain logistics to inhibit deterioration before the product reaches the consumer. A major challenge is the provision of small scale, reliable refrigeration units using hydrocarbon refrigerants to farmers in Asia and Africa.

The journey from farm to shop is the step in the food chain in which food is likely to be adulterated. As pressure on food supplies increases the incentive to adulterate will also increase. A number of commodities are commonly adulterated at the present time both at a local and at an international level. While it is rarely the case, as it was in Victorian England, that some commodities are scarcely ever procured pure, it is certainly true that many widely traded commodities are adulterated. The list includes milk, orange juice, vegetable oils, honey, herbs and expensive ingredients such as saffron, Manuka honey and ground pepper.

The success in the UK of an orange juice testing scheme has demonstrated that development of sophisticated analytical methods and regular testing can reduce the ease and profitability of adulteration to the point where it is no longer worth attempting to sell adulterated orange juice in quantity in the country. However, the lengths to which the fruit juice industry has had to go to produce this result have demonstrated the effort and investment needed. Currently, analytical techniques are extremely efficient at identifying very low levels of specific contaminants but the history of adulteration, for example the use of melamine to enhance the apparent protein content of milk, has shown that it is not easy to predict what will be used to adulterate commodities in the future. There is a challenge to develop techniques which will identify if the product is typical of its type rather than specifically what is different. It is already possible to identify protein, fat and water content of specific crops at the factory gate; it now needs to be possible in the same way to identify whether the product is typical of its type.

FOOD AND THE CONSUMER

The food industry is often discussed as if it were one homogenous entity but the challenges facing, for example, ready meals manufacturers are very different from those facing snack food manufacturers.

However, the overall challenge in the developed world is the same in all sectors – to provide food which is tasty and satisfying while being lower in calories, salt, fat and sugar at the same time reducing or, at worst, maintaining costs.

The food environment is changing. People in the developed world are spending less time in the kitchen and are buying meals that they expect to be able to prepare in minutes and which will be nutritious, satisfying and not obesogenic. In addition, eating out of the home is increasing. The most recent information from DEFRA [7] states that consumer expenditure on catering services was £87bn in 2016 compared to household expenditure on food and drink of £115.3bn. This chapter will consider some of the challenges facing the food manufacturing industry at the beginning of the 21st century.

SATIETY

The obesity problem, driven by the ready availability of cheap food and the decrease in physical activity, poses a number of problems. It is often said that people are living increasingly busy lives but this busyness is not related to physical activity. Being less active means people need fewer calories. The recommended daily intake for an active person is 2000 kilocalories for a woman and 2500 kilocalories for a man. It is a major challenge to provide sufficient protein, micronutrients and fibre in a diet containing less than this because the person is inactive. A related issue is the provision of tasty, nutrient dense, meals for the elderly who have lost their appetite.

However, we eat not only to gain the nutrients we need but also because it is pleasurable and we like to feel satisfied. The mechanisms of satiety are as yet poorly understood and research is needed. Because of the seriousness of the problem on a societal scale, research is needed simultaneously on both the mechanisms of satiety and on the development of products that might induce a feeling of fullness.

SUGAR REPLACEMENT

The current focus in the UK on the reduction of sugar in the diet illustrates some of the issues associated with reformulation. The reduction of salt programme in the UK has been relatively straightforward since, with the exception of preserved meats, salt is mainly used for flavour. However, sugar is a constituent of biscuits, cakes, confectionery, desserts and dairy products where in addition to sweetness it modulates flavour, provides body and mouthfeel, creates texture and assists in lowering the water activity of the product, reducing microbial growth and extending shelf life. The UK Government requirement is for free sugars (that is sugars added to a product rather than naturally contained within it) in food to be reduced by 20% [8]. So, for an existing successful brand, the object will be to produce a product that complies with the Government requirement but for the consumer not to notice a change or to find the product equally acceptable. When attempting to develop the required product the developer is faced with a number of issues. First they will be aware that the Regulation (1333/2008 EC) covering food additives means that not all additives are available for all types of food. Thus low intensity sweeteners, like maltitol and inositol, are permitted in cakes and confectionery but not in soft drinks. Intense sweeteners could be used, but the Regulation currently requires that intense sweeteners only be used in mixtures with sugars if the overall energy content of the product is at least 30% less than that of the original product. This reduction in use of sugar is greater than that required by the brief and will have other technical implications. The situation is not straightforward.

Some proposed ingredients have unwanted physiological effects in some consumers; inulin and other polyfructose polymers can cause intestinal problems for a proportion of the population and polyols can have laxative effects.

In terms of flavour, no sweetener has a sweetness profile identical to that of sugar and most have an aftertaste that makes the product less acceptable to some consumers. Reducing sugar in a cake recipe will impact crumb structure and product volume and even soft drinks are affected by the change in texture or "mouth feel" caused by removal of sugar; sugar free drinks would be perceived to be more watery. The shelf life of jam and preserves is generated by the high sugar content, reduce the sugar and it is necessary to use a preservative or keep the product chilled and change the label to declare the product as low sugar jam.

Inevitably other ingredients or additives have to be used to replace the product attributes of sugar. But this is in a consumer environment where a major trend is for simpler and "cleaner" labels and minimisation of ingredient lists and of complicated ingredient names. The complexity of this situation is driving research and development and some straightforward solutions, for example hydrolysing lactose (to glucose and galactose) in milk products to increase sweetness without increasing calorific content, have already been introduced. It will be a major challenge to see how far sugar reduction can be taken without significantly impacting consumer preference or product safety. The removal of an ingredient is rarely as simple as those outside the food industry would anticipate. This has already been amply demonstrated in the attempt to replace gluten in gluten free bread and rolls. While considerable improvements have been made to these gluten free products in the past 30 years, they are still very different to, and much more expensive than, the products they attempt to mimic

IMPROVING SHELF LIFE

The deterioration of a foodstuff in storage is affected by a combination of microbiological growth and chemical and enzymic reactions that occur within the food. The hurdles used to manage this deterioration produce a shelf life that allows the product to be processed, delivered and sold are unique to each foodstuff. They include reducing water activity (increasing the solids content of the foodstuff), adding a preservative, decreasing storage temperature, treatment to reduce microbial count or packing in an atmosphere which inhibits microbial growth. Because shelf lives of chilled foods generally are short, a matter of days, extending them can provide significant advantages in terms of reduction of wastage and while considerable progress has been made in understanding the details of the microbiology of complex mixtures of foodstuffs which underlie the shelf life of chilled foodstuffs more needs to be done to underpin developments in shelf life extension. In contrast, many packaged products are heat treated to reduce numbers of bacteria and provide a long shelf life. The microbiology of these treatments is well understood but they often have adverse effects on taste and texture. The development of high pressure processing has demonstrated that it is possible to reduce numbers of bacteria with little impact on taste but current high pressure systems are expensive and limited to few products. Irradiation has also been shown to be effective in a number of foodstuffs but its application is limited by legislation. These two processes demonstrate some of the issues faced by techniques currently being developed. It is worth noting that many significant advances in food preservation in the past, including the pasteurisation of milk, have initially been opposed by industry, retail and consumers. It is more important than ever that food scientists ensure that their developments are seen in a positive light by consumers before they are introduced commercially.

NEW PROTEIN SOURCES

As the world population increases the demand for protein will also increase. However, merely increasing numbers of animals will give rise to a number of problems. It is well established that animals are inefficient converters of feed protein into meat; chicken convert roughly 2kg feed into 1kg of meat, whereas beef cattle need anywhere from 4 kg to 7 kg feed to produce 1kg meat depending on where they are raised. Ruminant animals are also recognised as a significant contributor to global emissions of methane. It is clear that animal numbers cannot be increased to provide all the protein needed and new sources of protein will need to be developed. Vegetable and mycorrhizal proteins have been used for some time but are currently limited in their application, by both taste and texture. Further research and development is needed to improve the exploitation of these protein sources.

It has long been suggested that humans should eat more insects [9] and a number of insects are now being bred commercially, but on a relatively small scale. However, the insect flour is often merely being used as a protein supplement rather than as a useful ingredient in foodstuffs. Development is needed to improve the quality of insect protein rich foods, to identify the specific roles insect proteins can play as functional components of foodstuffs and to identify any adverse reactions to these proteins. Insect products are an example of a situation where marketing on the internet is ahead of legislation and larger scale introduction of the products in Europe will be dictated by compliance with the Regulation on Novel and Traditional foods from EFSA, which at present does not mention insects. Public reaction to eating insects is culturally dependent and, in some countries, it will be important to manage the introduction of products carefully and to improve the taste considerably from where we are now.

A further possibility is the use of leaf protein which is readily available but in low concentration. This would also be a novel food and, while it may represent an opportunity, many technical challenges in growth, harvesting, concentration and protection from oxidative degradation will need to be overcome and much evidence gathered on its safety before it could be used.

PROCESS CONTAMINANTS

We have recently become aware that processing foods, including by cooking, can give rise to unwanted chemicals. Acrylamide represents a problem initiated by inadequate knowledge and a desire to err of the side of safety (the precautionary principle). Since acrylamide was first identified in crisps, the food industry has made great efforts to reduce the amount in processed foods, but the evidence of harmful effects in humans resulting from exposure at the levels found in foods is still unclear. In a parallel issue, the soft drink industry changed from an effective preservative, sodium benzoate, to a less effective preservative, potassium sorbate, because it was found that, under certain circumstances, benzoate broke down to benzene, which is a known carcinogen. Whether the level of benzene in the soft drinks (being in parts per billion) was ever enough to be a risk to human health was never established, the suggestion of risk was enough. There are still specifications that require that the level of certain contaminants be zero. Lelieveld has recently commented on the appearance on antibiotics in meat arising from the presence of very low levels in soil as a result of the normal bacterial action [10]. However, as analytical techniques become more sophisticated, limits of detection are being reduced and more contaminants are being detected at low levels. Regulations will have to be better supported by science to establish finite limits; it is no longer acceptable to demand zero. Pesticide legislation, with its use of maximum residue limits, shows the way forward.

In addition, as analytical techniques become able to identify lower and lower levels of compounds, industry, academics and Government are going to have to become much more subtle in the assessment and communication of possible food safety issues and the use of the precautionary principle.

TRAINING

The food industry has made great strides in improving food safety in the past 50 years. The development and refinement of HACCP has enabled the provision of safe food around the world. However, the workforce in the industry has an ageing demographic profile. According to the US Bureau of Labor Statistics, the median age of employees in the food industry in the USA is 42.9 [11]. In the UK the Food and Drink Federation estimates that the industry will need a further 140,000 recruits by 2024 [12] It is important to train these people and to pass on the experience of those retiring to ensure that knowledge is not being lost. This is more than straightforward academic training. The modern apprenticeship approach in the UK may be a positive step because it involves significant time on the job learning from experienced team members but the adequate training of young staff is a major challenge for the industry. Training needs to be broadly based so that new members of the industry understand how problems in the past were solved and see how similar problems in other industries are managed. The Advanced Training Partnerships and their successor organisation, the AgriFood Training Partnership in the UK have gone some way towards providing ongoing training for food industry professionals in key areas and ensuring that skills are updated [13].

RESEARCH FUNDING

It would be remiss not to comment, in a chapter focussing on challenges faced by the food industry, on the funding of research in the UK and Europe at present. The approach over the past 30 years seems to have been based on the idea that without market orientation, researchers would behave as the residents of Laputa in Gulliver's travels and waste their time on meaningless research. The method chosen to avoid this is to require projects to identify their results before commencing and to run for short periods, requiring considerable expenditure of time in writing new grant applications (which may be rejected) in order to keep teams, who have developed skills, in place. It can take some 6 months for a chemist with a degree and some experience in analysis to become sufficiently familiar with a specific piece of equipment that they can produce reliable results worthy of publication. This training can then be wasted as the project could end in as little as 3 years. Some of the issues outlined in this chapter are long term and require long term approaches, rather than the inefficient short-term approach so common at present.

Conclusion

The food industry in the second decade of the 21st century faces significant challenges to continue to provide safe, wholesome food to an increasing population. As can be seen from this chapter, these challenges cover the entire food production chain from the field to the store. Providing solutions to these challenges will involve agencies from Government, trade bodies, research centres and individual companies and collaboration will be vital.

Many of these issues are dealt with in more detail in other chapters however it is important to realise the interconnectiveness of many of the needs. The food chain is complex and perturbations in one part inevitably have a knock on effect on other aspects. In addition the global nature of many of the challenges and the need to develop better ways of coping with diminishing natural resources will be a source of concern, pressure and innovation. The involvement of skills and sectors outside of the traditional areas of expertise has already been observed in some instances (e.g., the use of advanced satellite techniques in farming and agriculture) and it is likely there will need to be an increased number of such cross-disciplinary applications in the future. It is only by embracing the opportunities, with appropriate partnerships, that food safety and quality will be maintained.

References

[1] http://www.un.org/en/development/desa/news/population/2015-report.html.

[2] Webster, E; Cameron, R; Culham, A. *Gardening in a Changing Climate*. Royal Horticultural Society, 2017.

[3] Krebs, J. *Food: A Very Short Introduction*. Oxford University Press, 2013.

[4] Addressing the vulnerability of the global food system. *The Lancet*, Volume 390, Issue 10090, 95.

[5] Cordell, D; Drangerta, JO; White, S. The story of phosphorus: Global food security and food for thought, *Global Environmental Change*, 2009, 19(2), 292-305.

[6] http://www.fao.org/resources/infographics/infographics-details/en/c/317265.

[7] https://www.gov.uk/government/uploads/system/uploads/attachment_data/file/608426/foodpocketbook-2016report-rev-12apr17.pdf.

[8] https://www.gov.uk/government/uploads/system/uploads/attachment_data/file/604336/Sugar_reduction_achieving_the_20_.pdf.

[9] Holt, V. *Why not eat Insects*, Field and Tuer, Oxford University, 1885.

[10] Lelieveld, H. *AgroLife Scientific Journal*, vol 6 (1), 2017.

[11] https://www.bls.gov/cps/cpsaat18.htm.

[12] https://www.fdf.org.uk/publicgeneral/stats-2017.pdf.

[13] https://www.aftp.co.uk/

In: Food: The Vital Ingredient ISBN: 978-1-53614-402-4
Editors: A. B. Hanley and W. R. Russell

Chapter 5

FOOD ANALYSIS AND FOOD STANDARDS: RAW MATERIALS, TRACEABILITY AND AUTHENTICITY

Martin Rose[1] and Kathy Ridgway[2]
[1]University of Manchester, Manchester, UK
[2]Anatune Ltd, Girton, Cambridge, UK

FOOD ANALYSIS

Historical Context

The present-day Codex Alimentarius is the main international body responsible for food standards. The Latin name comes from the code used during the Austro-Hungarian Empire between 1897 and 1911 that was used to describe a variety of products and related standards. Food control however has a much longer origin and dates from ancient times. Past civilizations needed to protect consumers from dishonest and unsafe practices in food trade. Assyrian tablets of stone described the method to be used in determining the correct weights and measures for food grains, and Egyptian scrolls detailed the labelling required for certain foods. Beer and wines were inspected for purity and soundness in ancient Athens, and the Romans had a well-organized state food-control system to protect consumers from fraud or bad produce. In Europe, during the Middle Ages, individual countries passed laws concerning the quality and safety of eggs, sausages, cheese, beer, wine and bread. Some of these ancient laws still exist today (Rose et al. 2009).

When such factors are acknowledged by a society as important in order to assure fair play and to protect the consumer, there is an associated need for methods and procedures for enforcement.

The Role of Standard Methods

New technologies and analytical research enable us to measure new and emerging food contaminants and other chemicals that may be found to be a threat to the health of the consumer in the future based upon later discoveries or aspects such as structural analogy to existing toxic entities. Once risk assessment confirms the threat, legislation may be enacted to eliminate or limit the amount of these chemicals present in food that is sold. In order to enforce this legislation, it is necessary to be able to detect the presence of a chemical and measure its concentration in food. Hence, the consumer protection provided by enforcement depends on how well the measurement method performs. We need to be confident that measurement methods are performing sufficiently well to protect the consumer, without leading to the rejection of large quantities of food that comply with legislation. A number of approaches that can be used to provide confidence include: the use of standard methods, the use of analytical criteria that describe the performance of a method, and consideration of fitness for purpose based on measurement uncertainty (Horwitz, 1995; Rose, 2009a; Rose et al. 2011).

Food control laboratories are often generalist facilities that are required to offer a wide range of capability, often not only in the food sector, and while under variable and sometimes considerable financial restraint. Similarly the level of technology and analytical expertise may vary between laboratories. Hence, many such laboratories rely on the availability of standard methods from a variety of sources that are easy to implement, sufficiently robust and that have been shown to meet the requirements of legislation.

However, standard methods are not always the best approach to ensure food safety and public protection. The wide availability of standard methods which are designed, and demonstrated, to be reliably and economically implementable by a large number of laboratories is a valuable starting point for laboratories who need to produce valid measurement results in support of food safety and legislation. The use of standard methods will not be superseded as long as there remains a demand from non-specialist laboratories that need ready access to methodology that will be reliable and easy to implement 'off the shelf'. However, the use of a standard method is not always the best option for various reasons. The first reason is that, while we can be confident that standard methods are reliable and robust within their scope of application, the users of the method need to provide an objective measure that they are able to apply the method correctly to have confidence in the results produced by standard methods. The second reason is that technology and to an even greater extent 'events' move faster than standards. Standard methods will not always be available when we need them, or when they are available there may be other methods which could be more sensitive, accurate, cost-effective or rapid. We need to be able to tell when methods will produce results that are good enough to deal with a particular event (i.e., are fit for purpose), and when methods will produce results that are at least as good as a standard method (Rose et al. 2011).

The paragraph above mentions several reasons why we might need to know about the performance of an analytical method. And in general, 'users' of methods (and results produced by the methods) will want to know about method performance: analysts, risk managers, producers, consumers and legislators, all of whom might at some point need to understand what valid analytical methods can do. So, we need to have descriptions of the performance of valid methods of analysis that can be easily understood or translated.

Describing method performance using measurement uncertainty provides a versatile approach that encompasses, and makes comparable approaches based on formal validation by collaborative trial, the traditional parameters used by analysts to describe method performance and performance data generated during use of a method. Consideration of measurement uncertainty using approaches such as the uncertainty profile also makes it easy to assess and communicate the practical impact of uncertainty.

The main challenge for the analyst is that application of this approach relies more heavily on calculation and data analysis than on traditional multi-criteria laboratory validation. Largely this is a problem associated with the application of the bottom-up approach to uncertainty (estimating the size of each individual source of uncertainty and then combining mathematically), which may need many experiments undertaken specially to estimate uncertainty followed by many calculations and some algebra. However, in the last 10 years, simpler approaches to estimating uncertainty have become more popular.

Target Analysis and Non-Target Analysis

A foodstuff may contain one, two or multiple food chemicals that may be of concern for human health. Pesticides, veterinary drugs, natural toxins (mycotoxins, plant toxins) environmental contaminants, food contact materials and processing contaminants can and often do occur simultaneously. Each of these classes may have many hundred individual substances within it. Many of these chemicals are regulated and legal limits in food commodities have been established. For others, there are no limits, either because it is not established that they are found in food or because there is no evidence of human health concern. There is also a third class that falls between these categories of 'near-regulated' compounds. This includes chemicals where there is a known hazard, but regulations have not yet been established. Examples include new and emerging flame retardants, and compounds recently added to the Stockholm list of persistent organic pollutants. Where no legislative limit is set, or for the most hazardous compounds such as genotoxic carcinogens, methods must be able to measure to ensure that exposure is 'as low as reasonably practicable', which can be a challenge for analytical laboratories that do not have the very latest instrumentation. Limits may even be set before a suitable method has been identified (Mol, 2013).

An ideal system will measure a large number of known and unknown individual compounds, in a wide variety of sample types, at levels ranging from sub μg/kg to low mg/kg level. Currently, most regulated and near-regulated compounds are analysed using targeted methods, but with the development of time of flight MS systems, a non-target approach is becoming more widespread.

The traditional targeted approach relies on a knowledge of the compounds to be measured in order to develop methods and to perform analysis. A non-targeted approach conducts a complete or holistic analysis, followed by an extraction of the information required from the raw data. Non-targeted measurement uses a generic sample preparation that is as simple as possible to avoid the removal of compounds of potential interest. The simple extraction and clean-up required has the advantages that this will be cheap in terms of both equipment/reagents and also staff effort. This is followed by chromatography, combined with full-scan mass spectrometric detection, which is more expensive in terms of both equipment and staff time (and computer storage capacity). All ions are detected during the entire chromatographic run time and, in contrast to MS/MS, there are (in principle) no limitations to the number of substances that can be detected.

Using non-targeted measurement, it is possible to search raw data of previously measured samples for analytes that were not of interest or not known at the time of analysis, without re-sampling or re-analysis of stored samples, as long as data files have been kept. The output of each analysis generates a fingerprint, or total ion chromatogram (TIC), that can be used for comparison with previous or reference sample profiles to reveal deviations and to help with the identification of unknowns or to show differences between samples.

The potential of non-targeted measurement has been recognised for some decades with applications reported since the 1980s for toxicological screening of drugs. Soon after that, there were reports for the analysis of pesticides residues in food and of priority environmental pollutants in surface water. These early applications used GC-EI-MS (Gas chromatography-Electron Impact-Mass spectrometry, single quadrupole full scan) or GC-EI-ITD (ion trap detection), and were limited to compounds that could be analysed by GC. To improve selectivity, an increase in chromatographic resolving power was developed using comprehensive two-dimensional GC (GCxGC). This resulted in cleaner mass spectra and less interference of the diagnostic ions, and in an improvement in sensitivity. Due to the narrow peaks (~0.1 s), a clear advantage can be seen by the application of fast-scanning (>100 Hz) time-of-flight (TOF) mass spectrometry. The other possibility is to increase the resolving power of the mass spectrometer, which can result in an improvement in selectivity due to the fact that ions with the same nominal mass but with a different exact mass, can be separated, which is an alternative way to reduce interference with diagnostic ions. Instrumentation available until around a decade ago offered only intermediate resolution (5,000-7,000 FWHM) which meant that they suffered from limited dynamic range and unreliable mass accuracy. More recent instrumentation has better performance characteristics (13,000-25,000 FWHM).

The first high resolution TOF systems for LC presented similar limitations, but over the last decade, dynamic range, sensitivity and selectivity (resolving power/mass accuracy) have all improved. Resolving powers of up to 50,000 are common in modern instruments, and benchtop Orbitrap MS systems are capable of 140,000 (Mol, 2013).

Developments in HRMS detectors for LC have resulted in them being coupled to GC using alternative ion sources (APCI, atmospheric pressure chemical ionization). This soft ionization technique is somewhat less generic than EI, but the abundance of molecular or protonated ions combined with sensitivity and resolving power make it an attractive possibility.

Non-targeted measurements provide large quantities of data and storage can be a problem if all of the benefits of this technique are to be used – i.e., keeping files for retrospective examination. For each substance, not only analytes of interest, but all matrix compounds, information is gathered for retention time(s), accurate mass of ions (m/z 50-1000), isotope patterns and/or spectra, and intensity (from noise to saturation). It can be problematic to examine the data for all relevant information. Manual evaluation of spectra or extracted ion chromatograms takes too much time for routine application, and analyte detection has to be automated. This is done by using software and Libraries of spectra that are available, but in most cases, still needs analysts to check library assignments and in some cases chromatographic integration.

Validation of Non-Target Methods

Method validation of non-targeted methods needs to be approached in a very different way to conventional method validation. Conventional validation for veterinary drugs in food of animal origin and pesticides in food and feed in Europe is performed according to EU-guidance documents. A method is considered fit-for-purpose when a substance can be detected in 95% of the samples (with $N \geq 20$ in an initial validation exercise that needs to be complemented by on-going analytical quality control). Because even similar substances can behave very differently when analysed by chromatography-MS, the guidelines require validation to be performed at the individual analyte level. A major constraint for non-target methods is not so much sample analysis and validation, as having standards available for all of the compounds that might be found in a single analysis.

The variety and number of food matrices, and complexity of composition (liquid, solid, biological, fatty, acidic etc.) each provide a range of potential inferences. This can result in different approaches or methods for different food types for both targeted and untargeted analysis and make method validation a challenge (Mol et al. 2012).

FOOD QUALITY

The Perception of food quality is a critical element of consumer confidence in a product or brand. The colour, texture and flavour of food is used as an indication of quality and stability. Although 'Best before' dates used on food packaging indicate when a product may not be at its optimum, it may have no implication in terms of product safety.

Flavour Analysis

For flavour analysis it is possible to target analysis once the key aroma or taste compounds have been identified; however, for taints and off flavours a more investigative approach is required (Ridgway et al. 2010). For flavour matching or product development, a full characterisation study may be required.

For flavour analysis therefore there may also be several possible approaches. Since compounds responsible for flavour are often volatile, the vast majority of flavour analysis is done by gas chromatography, with either a Flame ionisation (FID) or mass spectrometric detector. Some more specific detectors, such as pFPDs for sulphur, or NPDs for nitrogen containing species may also be employed. There are some non-volatile components that are analysed by liquid chromatography, but these are more likely to be a handful of product specific target compounds. It should be noted, of course, that depending on the choice of sample preparation different profiles will be obtained. Those that offer some level of enrichment are most often employed (solid phase microextraction (SPME) stir bar sorptive extraction (SBSE), or dynamic headspace (DHS)), due to the limited sensitivity of static headspace.

Setting quality specifications for flavours – even once the key components have been identified, is rarely straightforward. Natural product variation, the synergistic effects of flavours and matrix components and stability issues all complicate analysis and sensory panels are still used extensively for quality control purposes and new product development. A good understanding of how analytical data relate to sensory analysis is still a challenge in the food industry. One technique that can help to bridge that gap is the use of GC-olfactometry (GC-O). In GC-O, the output from the GC column is split between two detectors – one of these being the human nose via an olfactory port. This approach enables odour active components in complex mixtures to be assessed by correlation with chromatographic peaks as they elute from the instrument (M. Brattoli et al. 2013).

As well as chromatographic techniques, more recently the food industry is using real time volatiles analysers to understand flavour release. This offers the advantage of being able to measure in-mouth and retronasal concentrations, as well as monitoring food production processes or critical compounds in a variety of matrices (Bowman and Barringer, 2012).

Flavour release profiles are strongly linked with consumer preference and, whilst the food industry has many mechanisms for prolonging or delaying flavour release, it is not always easy to measure analytically.

Techniques that provide real-time analysis of volatile organic compounds include ion molecular reaction mass spectrometry (IMR-MS), selected ion flow tube mass spectrometry (SIFT-MS) and proton transfer reaction mass spectrometry (PTRMS). All of these approaches use a soft ionisation process to reduce fragmentation, resulting in simple spectra that enable rapid monitoring of specific ions. Instruments using both quadrupole (unit mass) and accurate mass (time of flight) mass spectrometers are available. For SIFT-MS selectivity is achieved through a choice of reagent ions. Volatiles can be introduced to the system in a number of ways: directly (air sampling), from headspace vials, or from 'nose space' during consumption of a foodstuff.

Food Composition

Food composition analysis may be undertaken for several purposes – including labelling, authenticity, safety, nutritional claims, or product development. There are a plethora of techniques and methodologies used in food analysis from the robust and routine to the new, novel and developing. A number of tests are required to demonstrate due diligence in the food industry.

Nutrition – Nutrients and Vitamins

The analysis of nutrient content of food is underpinned by regulation and by analytical data built up over a number of years. The most common reference book for food compositional data is McCance and Widdowson's "The Composition of Foods." The data has now been collated into a database (https://www.gov.uk/government/publications/composition-of-foods-integrated-dataset-cofid) and is used as a reference by industry and academia.

Table 1. Chemical and Bioassay Based Methods

Analyte	Chemical Method	Bioassay Method
Vitamin A in foods and feeds	Multiple chromatographic steps followed by absorption at 325nm.	None
Thiamine (vitamin B1)	Chromatography, enzyme reaction to remove phosphate groups, chromatography, detection using thiochrome reaction	Growth of rats
Riboflavin (vitamin B2)	Fluorimetry	Lactobacilus casei growth curve
Niacin	Reaction with cyanogen bromide and absorbance after 1.5 minutes at 430nm.	Lactobacillus plantarum growth curve
Vitamin C	Reaction with indophenol and colour generation	Guinea pig growth bioassay
Vitamin D	None	Prevention of rickets in rats
Vitamin B12	None	Lactobacillus leichmanii growth curve

In the past many methods for vitamin and mineral analysis were both tedious and time consuming. For example, in the 1950s, prior to the widespread development and use of chromatographic techniques, a range of colorimetric and other methods were commonly used. For example, in the Official Methods of Analysis of the Association of Official Agricultural

Chemists published in the USA in 1960. Analytical methods for vitamins were divided into chemical and bioassay based methods. Some of these are summarised in Table 1.

More recently methods for multi-vitamin analysis have greatly increased the speed (and substantially reduced the cost) of analysis for vitamins in foods and feeds. In some cases, microbiological methods are still approved for use but these have been largely supplanted with rapid chromatographic and antibody based analytical techniques such as surface plasmon resonance. Nonetheless significant challenges remain for the analytical chemist including the effects of other components (interferences) in the system, sample preparation and speed of analysis. These will remain both obstacles and will also create opportunities for newer and better methods to be developed.

FOOD AUTHENTICITY

Authenticity of food is a general term that covers quality aspects to ensure that the product is as stated. This can cover aspects such as (a) provenance or geographic origin (is it from where it says it's from) and (b) Food adulteration (has this food been diluted or tampered with in some other way for (usually) economic gain). A themed collection of scientific publications showcasing the latest discoveries and developments in detecting food authenticity and integrity; including the analysis and detection of food fraud, contamination, adulteration and spoilage from the journals 'Analyst' and 'Analytical Methods' has recently been published (Goodacre and Ellis, 2013).

Determination of food authenticity is an important issue in quality control and food safety. Food authenticity testing is key for the food industry who are seeking to assure their food products labelling compliance and protect brands. As fraudsters are developing more sophisticated ways of adulterating food, analytical methods need to be able to identify whether the food is genuine/as labelled. Modern analytical techniques are capable of identifying known contaminants at very low levels but a more holistic analysis of food in its entirety is more of a challenge. The horse meat scandal was not initially picked up, as no-one was testing for horse (Brooks et al. 2017, and references contained therein).

Historical food fraud may have been easier to detect, where bulk ingredients were substituted or gross adulteration occurred, however, for some complex products, or where genuine products are diluted, analysis can be more of a challenge. It may be necessary to detect very low levels of an 'unknown' analytes to distinguish the fake from the genuine.

Many approaches to food authenticity use direct analysis with little or no sample preparation. However, in some instances some sample preparation is required to enable enrichment of low level analytes and look at the composition of a sample in more detail. The choice of sample preparation can be critical in the amount of information obtained about a sample, but can be challenging when the analytes of interest are unknown, or when only a small amount of sample is available. As counterfeiters are developing more sophisticated ways of adulterating food, analytical chemists have to look much harder at sample composition in order to combat food fraud and enable the protection of the food industry and the consumer.

Once identified, targeted analysis for known marker compounds can provide a rapid method for identification of authentic/non-authentic samples. The use of surrogate markers - for example measure of protein content for milk via Kjeldahl or dry matter content, have to be treated with caution (as can be learnt from the melamine in milk incident in China).

As well as chromatographic techniques, such as those previously described, spectroscopic approaches, such as NMR are being increasingly employed (G. Downey (Ed), 2016).

When using non-targeted/fingerprint approaches for natural ingredients or complex products it is critical for the food industry to understand what is within the normal variation of the genuine product (seasonal, annual, variety, location). Testing for authenticity of such foods relies on large data sets and statistical analysis in order to determine the 'odd one out' (Riedl et al. 2015).

Currently there is a lack of authentic databases and standardised testing protocols.

Similar analytical tools can also be used for purposes of Food Traceability (especially important for composite foods -can we tell the origin and producers for each component).

FOOD SAFETY

Microbiology

The microbiological safety of food is a major concern both to the consumer and to regulatory authorities. At line, on line and near line testing are challenging. In the UK the Advisory Committee on the Microbiological Safety of Food provides advice in response to requests from the Food Standards Agency and also on matters that Committee members themselves identify as important. It consists of independent experts drawn from a wide range of interests. It has produced a number of comprehensive scientific reports, prepared by Working Groups and based on the most up-to-date information available.

From a regulatory standpoint, EU Regulation No. 2073/2005 (amended by EU Regulation No. 1441/2007) on microbiological criteria for foods applies to all businesses which produce or handle foods. The regulation does not make end product testing mandatory or impose positive release criteria. Food safety in general and microbiological safety is a non-negotiable aspect of food businesses and trade organisations such as the British Retail Consortium and the Chilled Food Association have produced guidance on ensuring compliance with the regulation.

While many companies still employ in house microbiologists and will have a regular testing programme, there has been a rapid increase in the use of contract research organisations (CROs) who specialise in rapid testing and sample turn around. There are savings due to scale and specialism.

Testing is carried for pathogens, spoilage organisms and indicator organisms (those that, although not a problem in themselves, might indicate that there is a potential or hidden problem in the production chain). Analytical methods for microorganisms include classical culture methods, often using specific growing conditions and methods based on detection of DNA sequences specific to a given organism. In most cases it is necessary to amplify the amount of the DNA present using polymerase chain reaction (PCR) or a similar technology. Antibodies and chemiluminescent technologies are also commonly used for the rapid detection of food borne bacteria.

In addition to end product testing, challenge testing is a common procedure. Most food systems are not sterile and the detection of even a pathogen may not indicate that the product poses a risk to humans. The organisms have to be present in sufficient numbers to cause harm. One way of predicting the growth characteristics of a microbe in a specific food is to carry out challenge testing. In a challenge test a food is inoculated with relevant organisms. The product is then stored and tested for these organisms during shelf life.

Chemical (Residues and Contaminants)

Food contaminants cover a diverse range of compounds from a number of potential origins. They can be present at trace and ultra-trace levels and therefore identification and quantification in foods can present a challenge to the analytical chemist. Many known contaminants have levels of toxicological concern, or a maximum level permitted by legislation, but for banned substances, the analytical method must be capable of measuring as low as is reasonably practicable to demonstrate absence. Moreover, for unknown contaminants, such as those responsible for taints and off flavours, the method must also provide unequivocal identification of the causative compound(s).

Unwanted trace chemicals in food fall into two broad classes, residues and contaminants (Schrenk, 2017). Residues are from chemicals that are used in food production, such as pesticides or veterinary medicines, whereas contaminants are from either anthropogenic sources such as pollution or chemical by-products. Due to the variety of analytes that can be termed as either 'food contaminants' or 'residues', including pesticides, packaging contaminants, environmental contaminants, processing contaminants, mycotoxins and veterinary drugs, several methods of extraction are generally required. The choice of sample preparation techniques employed will depend on the contaminant(s) of interest, the food matrix and the final analytical determination/instrumentation to be used (Ridgway et al. 2007).

Requirements of the analysis must always be considered and the method used should be fit for purpose – in terms identification of unknown analytes, accuracy required (qualitative or quantitative) and sensitivity if working to known limits.

As in any food analysis, taking a representative sample is critical in obtaining meaningful results. In some cases, the compound responsible may not be homogenously distributed throughout the food ingredient or product, particularly if it is present as a result of external contamination or migration from/through packaging or for example around the injection site for a veterinary medicine used on a farm animal.

For the majority of food contaminants, the final analysis is performed using a separation technique, typically chromatographic, combined with an appropriate detector. The increased use of mass spectrometric detectors has enabled selectivity and sensitivity to be optimised at the detection stage and has led to the development of methods with minimal sample preparation (so-called 'dilute and shoot'). Such approaches provide rapid methods and can cover a wide range of analytes, with crude extraction (although minimal sample preparation can result in increased instrument maintenance).

Developments in direct ambient ionisation mass spectrometry techniques have led to methods with no sample preparation, or chromatography at all and applications in food analysis are increasing. These include direct analysis in real time (DART), desorption electrospray ionisation (DESI) and atmospheric solids analysis probe (ASAP). These recent innovations provide the ability to record mass spectra on samples with minimal sample preparation. Direct techniques such as these are useful for developing rapid screening approaches. Current applications for contaminants in food are limited, although such approaches have been reported for determination of melamine (Dane and Cody 2010) and vegetable oil authenticity (Ramos Catharino et al. 2005).

Biological (Allergens and GMOs)

Analysis for genetically modified materials in foods presents significant challenges. While inserted DNA is relatively easy to detect in raw materials (so long as the sequence of the inserted gene is known) using polymerase chain reaction (PCR) technology. The essence of PCR is that two small pieces of single strand DNA that are complimentary to the inserted DNA (primers) are used to distinguish the target DNA and to amplify it many millions of times so as to make it detectable. While raw food and ingredients are relatively easy to use, processing often makes DNA difficult to extract and, in many cases, it is so substantially degraded as to be undetectable using amplification techniques.

Different types of PCR are commonly used including quantitative, real time PCR to determine the levels of a given gene. Each of these methods provides both speed and specificity. There have been some advances in the detection of unknown sequences by using common characteristics of many inserts (e.g., the 35S promoter) however this is less of a problem in a tightly regulated market.

An alternative to the use of PCR is immune analysis. In this case the target is a protein rather than the DNA. One method makes use of antibodies in a dip stick type format. This method is generally qualitative (or semi-quantitative) and does not require specialised laboratory equipment therefore it is suitable for field testing. The second method – the enzyme-linked immunosorbest assay (ELISA) uses the same basic principle as the dip stick – the binding of a target protein to antibodies bound onto a solid surface – but in this case a multi-well plate is used and the detection method generally requires the use of a spectrophotometric or similar detector. The type of assay selected will depend both on where the test needs to be carried out (laboratory or in the field/factory) and the required level of sensitivity.

In the case of food allergens, common detection methods include antibody tests (ELISA), PCR for allergen DNA and, in some cases, chemical identification of proteins. The most common methods are those based around ELISA type formats. Once again, allergen detection in raw, unprocessed material is more straightforward than in processed material. In addition, processing can change the allergen profile of a food or food ingredient. Processing can also change the availability of allergenic parts of proteins (epitopes) and make them harder to extract – although they may still be present and be released after gastric digestion. In both processed and unprocessed samples, using antibody (epitope detection based) methods, there is a possibility of false positives and false negatives and difficulties in quantification. Recent advances have allowed LC-MS/MS methods (generally preceded by protein digestion) to produce a peptide map from a sample that can then be subjected to a database search and the unique map for the target allergen to be detected. Such methods can detect a number of allergens simultaneously. In both cases it is essential to have a clear paper trail of the history of a food product to try to prevent contamination.

The use of Blockchain methods has attracted some recent attention and a collaboration was recently announced between IBM, Nestle and Walmart. In principle, using Blockchain it will be possible to quickly identify the source of a contamination issue. The use of Blockchain will quickly link food producers, processors and the retail outlets in a single 'ecosystem' however it relies on cooperation between the different components of the supply chain and participants need to have '"permissioned access" to known and trusted information regarding the origin and state of food for their transactions'. The use of Blockchain may mark the end of the 'one up, one down' system of traceability in which each operator is responsible only for the stage of the chain under their control. By linking each stage in the chain whole chain traceability can, in theory, be achieved however there have been instances where this limited traceability information provision has been insufficient to prevent contamination issues arising.

FOOD STANDARDS

Risk Assessment

Chemical risk assessment provides the scientific basis for decisions aimed at ensuring, maintaining and improving the safety of human exposure to chemicals. The risk assessments that underpin development of regulatory measures for chemical contaminants and residues in food are generally conducted by authoritative independent committees of scientific experts, such as the scientific panels of the European Food Safety Authority (EFSA), and the bodies that advise the Food and Agriculture Organization (FAO) and World Health Organization (WHO) of the United Nations, i.e., the Joint FAO/WHO Expert Committee on Food Additives (JECFA) and the Joint FAO/WHO Meeting on Pesticide Residues (JMPR).

Risk assessment is defined as 'a process intended to calculate or estimate the risk to a given target organism, system or (sub)population, including the identification of attendant uncertainties, following exposure to a particular agent, taking into account the inherent characteristics of the agent of concern as well as the characteristics of the specific target system' (IPCS). In the context of chemicals in food, the term safety assessment is also sometimes used, wherein safety is the 'practical certainty that adverse effects will not result from exposure to an agent under defined circumstances' (IPCS). Risk is defined as the 'probability of an adverse effect in an organism, system or (sub)population caused under specified circumstances by exposure to an agent' (IPCS).

Since estimation of exposure is such a key part of the risk assessment process, it is clear that reliable analytical methods are paramount (Alexander et al. 2012; Benford, 2013).

Regulation

Different global regions have different approaches to regulation, legislation and enforcement. The degree and scope of food safety regulation correlates strongly with socio-economic development: In least-developed, very poor regions, ensuring an adequate food supply in order to prevent starvation is the highest priority, and less attention is given to good agricultural practice. Food safety controls in these countries may be practically non-existent.

Within the EU, food is regulated at the EU level, but individual countries can also have specific regulations where there are specific regional needs or to cover specific local consumption or production practise etc. Other industrialised countries such as US, Canada, Australia, New Zealand and Japan also have highly evolved systems of food regulation. In some less-developed countries, food control can be more focussed on requirements for export to more developed countries and the need to meet regulations in force in the export country or region. This can lead to a two-tier system with more rigorous control on food for export than for the local market.

In addition to official control undertaken in the public sector, there is increasing reliance on industry for self-regulation and due diligence measures to ensure that unsafe products or foods that do not comply with regulations are not sold. Producers are also keen to portray the quality aspects of their products. A number of approaches exist to support these objectives.

Many producers apply standards such as HACCP or GMP to give confidence in production methods. Major retailers can enforce their own strict standards on their suppliers.

Trades associations can develop their own quality marks such as the UK 'red tractor', soil association standards for organic produce, British Lion for egg production etc.

Guaranteed provenance for PDO goods certifies the origin of protected regional products such as Parmesan Cheese, regional wines, Feta cheese etc. In order to implement all of these standards, it is important that analytical methods are available to detect fraudulent use of these standards, and the main drive for these is from industry in order to protect brands.

National and international authorities are seen as key stakeholders because of their mandates to protect consumer safety and ensure fair trade and consumer confidence. Within the European Union (EU), a network of European Reference Laboratories (EURLs) and National Reference Laboratories (NRLs) was set up to assist the work of public authorities. Their role includes work on the harmonization of food control across the EU. NRLs work closely with the Central EURLs appointed by the European Commission and with official national control laboratories within their own country. The EURL and NRLs also provide a channel for communication among relevant authorities and provide input for the production of standard operating procedures, codes of practice and guidance documents. They also provide input about compliance audits, organize interlaboratory comparisons and arbitrate in cases of dispute. Further rapid responsiveness and communication is facilitated via the Rapid Alert System for Food and Feed (RASFF), in place since 1979. Keeping its structure simple with clearly identified contact points and information exchange in clear and structured formats ensures the effectiveness of this alerting and information-exchange mechanism.

The EURLs, NRLs and RASFF together have a key operational role in the development, implementation and communication of harmonized monitoring of food contaminants. Food regulations are subject to regular updates, and it is important to always check that the latest version is the one that is being used to check compliance.

Acknowledgment

The authors wish to thank Bryan Hanley for contributing sections of text on some of the biological aspects discussed in this Chapter.

References

Alexander J, Benford D, Boobis A, Eskola M, Fink-Gremmels J, Fürst P, Heppner C, Schlatter J, van Leeuwen R; (2012) Special Issue: Risk assessment of contaminants in food and feed. *EFSA Journal* 2012;10(10):s1004. [12 pp.]. doi:10.2903/j.efsa.2012.s1004. Available online: www.efsa.europa.eu/efsajournal.

Benford DJ (2013) Risk assessment of chemical contaminants and residues in foods. In *Persistent organic pollutants and toxic metals in foods*

Edited by Martin Rose, Alwyn Fernandes. Published by Woodhead publishing Ltd, Cambridge, UK. ISBN 0 85709 245 6.

Bowman and Barringer, 2012, Analysis of factors affecting volatile compound formation in roasted pumpkin seeds with selected ion flow tube-mass spectrometry (SIFT-MS) and sensory analysis. *J Food Sci.* 2012 Jan;77(1):C51-60.

Brattoli M, et al. (2013) Gas Chromatography Analysis with Olfactometric Detection (GC-O) as a Useful Methodology for Chemical Characterization of Odorous Compounds; *Sensors* (Basel). 2013 Dec; 13(12): 16759–16800.

Brooks, Stephanie, Elliott, Christopher T., Spence, Michelle, Walsh, Christine, Dean, Moira. (2017). Four years post-horsegate: an update of measures and actions put in place following the horsemeat incident of 2013. *NPJ Science of Food.* 1:5 1-7; doi:10.1038/s41538-017-0007-z.

Catharino Ramos, et al. 2005, Characterization of Vegetable Oils by Electrospray Ionization Mass Spectrometry Fingerprinting: Classification, Quality, Adulteration, and Aging, *Anal. Chem.,* 2005, 77 (22), pp 7429–7433.

Dane and Cody 2010, Selective ionization of melamine in powdered milk by using argon direct analysis in real time (DART) mass spectrometry. *Analyst*. 2010 Apr;135(4):696-9. 2010 Feb 5.

Downey G, (Ed), 2016, Advances in Nuclear Magnetic Resonance Spectroscopy for Food Authenticity Testing, in book: In book: *Advances in Food Authenticity Testing,* 2016, Woodhead Publishing, pp.147-170.

Goodacre and Ellis (2013). Detecting food authenticity and integrity. *Anal. Methods,* 2016, 8, 3281-3283 http://dx.doi.org/10.1039/C6AY90015A

Horwitz W. (1995) Protocol for the design, conduct and interpretation of method performance studies. *Pure and Applied Chemistry*, 67, 331–343.

Mol Hans, (2013). Non-target is our target. *The Analytical Scientist*, May 2013 pp 24-29.

Mol HGJ (2012) Reynolds SL, Fussell RJ, and Štajnbaherc D. Guidelines for the validation of qualitative multi-residue methods used to detect pesticides in food, *Drug Test. Analysis* 2012, 4 (Suppl. 1), 10–16.

Ridgway et al. (2007), Sample preparation techniques for the determination of trace residues and contaminants in foods, *Journal of Chromatography A,* Volume 1153, Issues 1–2, 15 June 2007, Pages 36-53).

Ridgway et al. (2010), Analysis of food taints and off-flavours: a review, *Food Addit Contam Part A* 2010 Feb;27(2):146-68.

Riedl et al. 2015, Review of validation and reporting of non-targeted fingerprinting approaches for food authentication, *Analytica Chimica Acta,* Vol 885, 2015 17-32.

Rose Martin. (2009a) Research methods, analytical criteria and standard methods: help or hinderence? pp 81 in *-Recent advances in food analysis*; Edited by Jana Pulkrabová, Marie Suchanová and Monika Tomaniocá. Published by the International Association of Environmental Analytical Chemistry ISBN 978-80-7080-726-2.

Rose, Martin, Poms, Roland, Macarthur, Roy, Pöpping, Bert, and Ulberth Franz, (2011) What is the best way to ensure that valid analytical methods are used for food control? *Quality Assurance and Safety of Crops and Foods* 3 123-134.

Rose, Martin, Thomson, Barbara, Jensen, Anne-Mette, Giorgi Liana, and Schulz Claudia. (2009) Food monitoring and control for environmental contaminants. *Quality Assurance and Safety of Crops and Foods*. 1 (3) 160-169.

Schrenk (2017) (Editor). *Chemical contaminants and residues in foods*. Editor: Professor Dieter Schrenk, Technical University of Kaiserslautern, Germany. 2nd edition. 29 June 2017. Pub: Woodhead (Elsevier) eBook ISBN: 9780081006757 Hardcover ISBN: 9780081006740.

In: Food: The Vital Ingredient ISBN: 978-1-53614-402-4
Editors: A. B. Hanley and W. R. Russell

Chapter 6

SUSTAINABILITY IN A GLOBAL AND LOCAL CONTEXT

Liliya Serazetdinova, PhD*
and Jayne L. Brookman, PhD
Knowledge Transfer Network, London, UK

INTRODUCTION, ISSUES AND BIG CHALLENGES

The global agri-food industry currently faces a multitude of challenges, including growing demand for affordable food, decreasing land availability, changes in weather patterns and diseases, increasing production costs, volatile food prices and uncertain profit margins [1]. Consumer demand for healthy and fresh food is on the increase and regulatory requirements must be met across the supply chain.

Food and drink producers relying on bulk commodities such as wheat, fruit, vegetables, oils and fats, are particularly affected by the need to source

* Corresponding Author: liliya.serazetdinova@ktn-uk.org

environmentally sustainable and resilient raw materials, and rely on technological innovations to meet this challenge.

Impact of Climate Change on Food Supply

In terms of its effect on agriculture and food production, climate change presents a great challenge both today and in the immediate future. Between 30% – 35% of global greenhouse gas (GHG) emissions come from agriculture [1].

In addition, the food chain also has an impact on greenhouse gas emissions. In Europe alone, food accounts for 31% of the EU-25's total GHG impacts, with a further 9% arising from the hotel and restaurants sector [2]. Understanding, measuring, controlling and reducing emissions from within the food supply-chain requires chemical, biochemical and chemical engineering solutions.

Carbon Footprint of Food Production

Food production leaves a footprint on the environment associated with use of resources throughout their production, use and disposal. This footprint is associated with carbon emissions and use of water throughout all stages of the agri-food supply chain, from production of agrochemicals, use of agricultural machinery, transportation of raw materials, processing, food manufacture, retail and disposal of agricultural and food waste. In order to assess all environmental impacts of food production, or indeed any product or a process, a technique is used called life cycle analysis (LCA). LCA measures all environmental impacts associated with production, use, and disposal of a product, from use of land, sea and air, water, chemicals and energy. LCA is used to compile an inventory of relevant energy and material inputs and environmental releases, evaluate the potential impacts associated with identified inputs and releases, and interpret the results [3]. LCA allows for comparison of products in relation to their environmental impacts, including carbon emissions (Global Warming Potential), contamination of air, land, and water courses, water use, land use change, euthrophication, and effects on biodiversity. LCA can be used to optimize product or process design, to identify ways to save energy and other resources.

Although very powerful, LCA has its limitations. It is not always possible to have accurate data and practitioners often rely upon generic data, averages, or even assumptions based on other data. With the lack of consistency in the methods and assumptions, comparative LCAs of similar products could sometimes lead to different and potentially contrary conclusions. It is also very difficult to account for social implications or to measure impacts.

Major food manufacturing companies all over the world use LCA to evaluate environmental impacts of their products. Third-party certified LCA-based labels provide an increasingly important basis for assessing the environmental impact of competing products, and support marketing efforst to promote safer and environmental friendlier products to customers.

Displacement of Biodiversity and Change of Land Use and Effect of Pollinators

The current world population of 7.6 billion is expected to reach 8.6 billion by 2030, 9.8 billion in 2050 and 11.2 billion in 2100, according to a new UN report [4]. Food production (including crops and livestock) currently occupies ca. 40% of ice-free land (4,300 million ha) [5]. Cultivated areas may need to grow additionally by 2.7–4.9 million ha annually to satisfy growing food demands due to rising world population [6, 7].

Crop and livestock production fundamentally change how natural environment is used and affects the function of natural ecosystems, causing changes on biodiversity, geological carbon cycle, and water purification, and change of habitats for living organisms. Conversion of land for agriculture is estimated to account for 80% of global deforestation [8, 9], and ca. 53% of threatened terrestrial species are negatively impacted by agriculture [10]. These changes also affect the amount of carbon stored in the environment, with 24% and 10% less carbon is stored in vegetation and soil under today's agricultural activity, respectively, compared with landscapes retained their natural vegetation [11].

The use of synthetic nitrogen fertiliser has increased nearly 21-fold since 1950, and more nitrogen is now added to agricultural soils than from natural processes, and most human-derived nitrogen is lost to the atmosphere or receiving waters [12]. Agricultural waters run-off cause eutrophication of waterways, which negatively impacts aquatic ecosystems [13]. Phosphorus from fertiliser and livestock manure, pesticides, and nanoparticles are also exacerbating environmental pollution [14]. Use of synthetic pesticides was one of the main factors underpinning increasing crop yields and security of food supply over the last half of century (see Chapter 11), however, their increased use leaves significant negative impact on water, soil, and living organisms, especially pollinating insects. With pollinators required for 30% of the human food supply [15], negative impacts on bee populations could lead to decreased reliability and diversity in food production. Among other factors, use of pesticides and other agrochemicals is associated with decline in pollinator populations. The prevailing evidence from scientific and regulatory community is that the general declining health of honey bees is related to complex interactions among multiple stressors including pests and diseases, poor nutrition associated with loss of foraging habitat and increased reliance on supplemental diets, pesticide exposure, bee management practices, and lack of genetic diversity [16].

Water Availability

Food production largely relies on use of water, from crop production to food processing. Climate change has serious impacts on water systems around the globe, with more flooding and droughts. Freshwater supply around the world are already under excessive pressure from drainage, dredging, damming, pollution, extraction, silting and invasive species. Some of the driest continents managed to sustain agriculture, food production and urban life largely due to availability of groundwater reserves, however these are being depleted all over the world. One of the sources of ground water is an aquifier, a layer of porous substrate that contains and transmits groundwater. Aquifiers are present on all the Earth's continents, and by the beginning of the 21st century, a third of the world depended on aquifers for drinking water and farming [17]. New modeling of the world's groundwater levels suggests that aquifers around the world are getting depleted. By 2050, as many as 1.8 billion people could live in areas where groundwater levels are fully or nearly depleted because of excessive pumping of groundwater for drinking and agriculture [18]. According to a recent study linking ground water deletion to international food trade, rice used 29 percent of the groundwater, followed by wheat (12 percent), cotton (11 percent), maize (4 percent), citrus (5 percent), sugar crops (5 percent), and soybeans (3 percent) [19]. Changing weather patterns will require shifts in agricultural practice and food processing, including improving water deficit resilience in crops, increasing yield, reducing water pollution, and novel water-free processing technologies. These improvements are underpinned by chemical and biological sciences.

Agricultural Residues and Food Waste

Estimated total UK post farm-gate food and drink waste is around 10 million tonnes per year, 60% of which could be avoided [20]. This has a value of over £17 billion a year, and is associated with around 20 million tonnes of greenhouse gas (GHG) emissions. Levels of food and drink waste by commercial and industrial businesses in the food sector amount to around 2.8 million tonnes, with a value of over £4.3 billion [21]. At least 14m tonnes of bio-based residues are produced in the UK from crops and forestry sources each year [22]. Examples of agri-food wastes that can be recovered include nitrogen and phosphorous run off, arable crop residues, forestry waste, potato starch, animal manures and slurries, whey from milk production, AD sludge, biogenic fraction of municipal soild waste (MSW), methane emissions from ruminant livestock, animal by products, animal bedding, poultry litter, surplus crop and vegetable production, vegetable peelings and vegetable rejects, in storage losses, seafood and fishery waste, agricultural packaging, and agricultural machinery waste oil. A range of chemical and other biological techniques can be used to optimise recovery and re-utilisation of these waste materials.

Soil Health

Extensive use of agrochemicals and heavy agricultural machinery leads to soil erosion. Soil health is being recognised as one of the key challenges in the agricultural sector. There is an urgent need to improve soil health across a range of agricultural systems including arable rotations, grassland systems, and perennials. There has been a surge of international and national efforts to better understand and manage soil health. FAO in its position paper "Agriculture and soil biodiversity" [23] urges the agricultural sector to put more emphasis on improving soil biodiversity, and promotes sustainable soils management and agro-ecological farming. In the UK, there has been an extensive recent effort to understand our knowledge of soil health, and review policies that would help to manage it. In 2015, the Department of Environment and Rural Affairs (DEFRA) launched a Soils Evidence Review [24] that identified key research priorities for developing a better understanding soil functions. The UK Parliament launched in 2015 a Soil Health Enquiry [25], the outcomes of which stated, that some of the most productive agricultural land in England is at risk of becoming unprofitable within a generation through soil erosion and loss of carbon, and the natural environment will be seriously harmed, and called for more action. Research Councils, with contributions from DEFRA and Scottish Government, are funding a Soil Security Programme [26], aiming to improve our understanding of soil functions, and how soil reacts to climate change. The Agriculture and Horticulture Development Board in the UK (AHDB) funded a number of research initiatives that are designed to develop practical tools for farmers [27]. While a lot of knowledge and tools have been recently developed, there is an urgent need to integrate these into conventional agricultural practices. A number of approaches could be deployed, including use of cover crops and crop rotation, soil amendment products to improve soil biodiversity, and structure, and better farm management planning to reduce soil compaction.

INITIATIVES, SOLUTIONS, AND ROLE OF CHEMICAL SCIENCES

Precision Agriculture

Precision agriculture is a farm management approach that takes into account variations in the field (or on a livestock farm) to produce more food using fewer resources and reducing production costs. Every aspect of the production system – soil, weather, vegetation, water – vary from place to place. And all these factors determine crop growth and farming success. Precision farming maps and manages these variations precisely, and applies throughout all stages of crop production, from soil preparation and seeding, to crop management and harvesting. In livestock farming, precision agriculture relies on automatic monitoring of individual animals and for monitoring animal behavior, welfare and productivity, and their physical environment. Precise monitoring of these factors helps farmers to increase livestock production and quality, individually tailor animal care for in terms of feeding, milking and housing, and reduce farmers' workload, aiming to improve the overall sustainability and ensuring economic viability of farm operations.

The implementation of precision farming has become possible thanks to the development of sensor technologies, positioning systems, such as Global Navigation Satellite Systems (GNSS) and Controlled Traffic Farming (CTF), auto-guiding systems, and machine vision methods. These technologies help to link mapped variables to appropriate farming practices such as tillage, seeding, fertilization, herbicide and pesticide application, harvesting and animal husbandry. Measuring more precisely crop demands in fertilizer and crop protection chemicals, allows farmers to reduce application of agrochemicals, thus reducing pollution to groundwater and soil.

Innovations in the recent years are focused on precise and targeted application of agrochemicals, including improved formulation, delivery, and new chemistry with lower environmental impacts, or biological control methods. Integrated pest management approach (IPM) is used to take advantage of all appropriate pest management strategies, involves integrating multiple control methods including prevention by removing conditions that attract pests, as well as a combination of physical and chemical control methods.

Precision irrigation helps to address water shortages with technologies that use accurate indicators of water stress and help to apply variable rates of water depending on actual crop water demand, maximising the water use efficiency.

Precision farming can make a difference to food production facing the challenge of a rising world population and can help farmers to achieve greater sustainability, higher productivity, and realise economic benefits. Other benefits come from better working conditions, increased animal welfare and the potential to improve various aspects of environmental stewardship.

New Production Methods and Urban Farming

A key challenge for agriculture is to produce 70% more food by 2050. New crop production systems including indoor and soil-free growing can reduce pressures on land, chemical use, and support urban farming. The global market for hydroponics is projected to reach 395.2 Million USD by 2020[2].

Indoor and soil-free growing systems allow for better control of key environmental factors, such as water, light, heat, agrochemicals, and allow to significantly increase yield and reduce production cycles. Produce grown

[2] Markets and Markets, 2016

indoors use up to 98% less water and 70% less fertiliser than traditional farms [28]. Key innovations include the use of LED lighting, vertical growing systems, and soil-free growing. Soil-free growing can achieve higher yields as well as higher water and fertiliser use efficiency, compared with traditional agriculture, however usually has higher energy demands due to heating, cooling, lighting, and circulation requirements. Several soil-free growing methods are used including hydroponics, aquaponics, and aeroponics. In hydroponics, plant roots supported by inert medium are submerged into a nutrient solution. Hydroponics can use up to 70% less water than conventional irrigation-fed systems. Aquaponics is an integration of aquaculture (fish cultivation) and hydroponics into one production system. A symbiotic relationship is designed whereby waste from fish is used as a fertiliser for growing plants. In aeroponics, water is sprayed onto bare roots, rather than immersing them. It uses up to 70% less water than hydroponics [28].

Indoor and urban farming systems makes economic sense for perishable and hard to transport fresh produce, and high value crops with short production cycles. These production systems could help to deliver savings in production, transport, and packaging. Urban spaces including roof tops, underground tunnels, and data centres [29] are being increasingly used to produce food. Urban farming is driving the development of new business models and shorter supply chains. Examples of companies that have commercial solution for indoor and vertical crop production include GrowUp Urban Farms [30], Saturn Bioponics [31], and Growing Underground [32]. However, urban farming will not be able to scale up in a way that it can replace field grown produce, there might be structural and logistical reasons the production will stay limited. Despite these challenges, the development of indoor production is an important component in future food production as pressures on land and water use are growing, and these systems allow for significant savings of the resources, and improved environmental impact. New technologies and solutions, including plant feeds, biostimulants, crop protection agents, hydroponic substrates, as well as water and nutrient delivery systems, are being actively developed and tested.

Landscape Approaches to Food Production

Agricultural expansion and intensification change the way how natural environment is used. With rising demand for producing more food from the land, and increasing pressures on ecosystems and effects of climate change, managing landscapes in a sustainable manner becomes very important for the food production. Landscape approaches have developed from a need to identify integrated solutions to competing demands on land uses, pressures on environmental systems and requirements from the local community. Each landscape has a variety of uses such as agriculture, grazing, forestry, as well as recreational uses, ecosystem services such as water catchments and biodiversity habitats. Some landscapes provide less opportunity for large scale crop or livestock production.

An example of a specialized landscape is the uplands in the UK, which have limited potential for high value economic use, due to a combination of low accessibility and limited potential for vegetation growth. However, most upland areas in the UK provide a range of social and economic opportunities for people. Some of these include farming, forestry, military training, recreation, water supply, mining and quarrying. Yorkshire Dales in the UK is an upland area characterized by a combination of farming (sheep grazing on the uplands, and cattle on hay meadows), quarrying, and recreational use. Uplands play an important role in preventing or reducing the impacts of climate change. Land management which supports natural processes to increase the extent and improve the condition of woodland and natural vegetation will sequester more carbon, inhibit flooding, increase ecosystem functionality and support richer wildlife communities.

Another example is applying landscape approach to managing water supply risks by a drinks manufacturer. SABMiller, one of the world's largest brewers, recognised that water quality in their South African and Colombian breweries was a serious problem that needed to be addressed beyond their breweries. They turned to WWF and German Corporation for International Cooperation (GIZ) to develop a landscape approach aiming to mitigate risks, including planning and mapping of water supply, developing deeper cooperation between all the key players in the catchment, and financing of eco-system services. These landscape approaches have resulted in a reduction of operational risk and creation of reputational benefits for SABMiller, as well as significant benefits for other water users [33].

By focusing on a particular landscape, farmers, businesses, and communities may find opportunities for joint investments, utilisation of by-products or rotation planning to bring added value for farmer, local communities, and businesses. Landscape approach to food production links with principles of circular economy and bioeconomy, which seeks opportunities for recycling water and materials and is underpinned by chemical and biological techniologies for material recovery and recycling.

Private Sector Sustainability Standards

Underpinned by the need to reduce costs, improve quality and secure raw material supply, and the need to improve their brand value, private sector is implementing a number of sustainability standards. Examples of such standards include Organic, Fairtrade, Rainforest Alliance, 4C Association, and ProTerra. These standards and certification have become recently the dominant method of defining, delivering, demonstrating and creating demand for more sustainable products - particularly addressing issues in agriculture, fisheries and forestry. Estimated trade value of top 12 sustainability standards initiatives equals $31.6bn (USD). About two thirds of standards have a consumer facing claim. Standards-compliant production of major food crops has grown significantly, especially for certain commodities, such as bananas, coffee, palm oil, cotton, cocoa, soybeans, sugar, and tea, grown in Africa, South America, and Asia [28].

Despite the successful implementation, a number of challenges are hindering the current model of sustainability standards from delivering more widespread and transformative environmental and social outcomes, including evidence of impact, cost of implementation and complexity, data inaccessibility, and a shortage of agricultural training capacity needed to increase knowledge of better management practices. New approaches are beginning to be developed to address these challenges. These will use innovations in data collection as well as completely new models of measuring sustainability outcomes. Identifying opportunities to increase the efficiency of these standards, and in some cases finding completely new models of delivery, will be a key challenge for the food industry in the next ten years.

PUBLIC AND PRIVATE SECTOR INITIATIVES

The Courtauld Commitment

The Courtauld Commitment is a voluntary agreement aimed at improving resource efficiency and reducing waste within the UK grocery sector. The agreement is funded by English, Scottish, Welsh and Northern Ireland governments and delivered by WRAP. It supports the UK governments' policy goal of a 'zero waste economy' and climate change objectives to reduce greenhouse gas emissions. Since 2005, WRAP working in partnerships with retailers, brands and manufacturers, hospitality and food service, trade and sector orgnaisations, and local authorities, delivered a significant reduction in packaging and food waste, and a reduction in greenhouse gas emissions [34]. The Courtauld Commitment 2025 [35] is the new phase of this voluntary agreement aiming to achieve by 2025 a 20% reduction in food and drink waste arising in the UK, a 20% reduction in greenhouse gas intensity of food and drink consumed in the UK, and a reduction in impact associated with water use in the supply chain. The Commitment will drive best practice through its unique whole-system approach to the way food and drink is produced, sold and consumed in the UK.

M&S Plan A 2025

In January 2007, M&S launched Plan A, a five-year, 100-commitment, sustainable business programme [36]. Plan A was designed to address the key environmental, social and ethical challenges facing M&S, it was updated in 2010 and 2014, with new commitments to be achieved by 2015 and 2020 [37]. In January 2017, Marks & Spencer launched a new sustainability plan – Plan A 2025. Their ambitious, customer focused plan builds on the success of the first 10 years of Plan A and will support 1,000 communities, help 10 million people live happier, healthier lives and convert M&S into a zero-waste business. The three-pillar plan is designed to tackle the big issues facing retailers, consumer businesses and society. The commitments pledged in this plan include making all M&S packaging ‘widely recyclable’; raising £25 million for charities tackling cancer, heart disease, mental health problems, loneliness and dementia; aiming for 50% of food sales coming from healthier products; committing M&S workers to work-time community volunteering; ensuring all key raw materials M&S uses coming from sustainable sources; and a commitment to work with local councils and charity partners to support communities to deliver positive, measurable change.

Love Food Hate Waste

In 2007 WRAP launched the Love Food Hate Waste (LFHW) programme which helps UK households recognise and tackle food waste. Since its launch, millions of people have responded, saving around £1.5 billion worth of food [38]. Love Food Hate Waste is working with retailers and brands to help develop their own campaigns by providing resources such as well-researched tips, recipes, messages, and consumer insights. WRAP’s LFHW website [39] provides useful reference and communications material that local authorities can draw on to promote food reduction in their area. It contains ideas to help individuals, communities and organisations reduce food waste as well as the LFHW partner website [40] containing free materials, templates, and resources for local authorities. The campaign involves a wide range of partners, from community organisations, chefs, UK businesses, trade bodies and local authorities through to individuals. LFHW focuses on five key overarching behaviours: planning, portions, date labels, leftovers and forgotten foods and storage.

EU Circular Economy Package and Bio-Economy

Agriculture, forestry, fisheries and aquaculture, together with the bio-based industries, form bioeconomy. All these industries use limited natural resources, to satisfy the demand of consumers and a wide range of industries for food, feed, bio-energy and bio-based products. The bioeconomy interconnects many supply chains in agriculture, food, chemical, biotechnological, energy, and construction industries. The bioeconomy is based on the principles of biorefineries that use biomass, by- and co-products and waste instead of fossil resources as their raw material. In the bioeconomy, land use and food security are optimised through a sustainable, resource-efficient and largely waste-free utilisation of renewable raw materials. Estimates conclude that a shift to biological raw materials and biological processing methods could save up to 2.5 billion tons of CO2 equivalent per year by 2030, increasing markets for bio-based raw materials and new consumer products several-fold [41]. In the UK, bioeconomic activity is currently estimated to be worth £56 billion direct GVA and £164Bn is the GVA from activities upstream and downstream of the bioeconomy [42].

Bioeconomy is underlying circular economy related to industries utilising biological resources. Circular economy is a concept where production, use and disposal of products forms a loop, whereby waste or by-products generated during production and use are turned into a valuable resource again. Unlike in the traditional, linear economic model, based on a 'take-make-consume-throw away' pattern, a circular economy implies reducing waste to a minimum as well as re-using, repairing, refurbishing and recycling existing materials and products. Circular economy approach can help food producers to reduce pressures on the environment, enhance security of supply of raw materials, increase competitiveness and innovation. Both bioeconomy and circular economy heavily rely on chemical sciences to optimise the use of resources.

The European Commission adopted an ambitious Circular Economy Package [43], which includes revised legislative proposals on waste and consists of an EU Action Plan for the Circular Economy that covers measures related to production, consumption, waste management, and the market for secondary raw materials, proposing to close the loop of product lifecycles through greater recycling and re-use, and bring benefits for both the environment and the economy, including the revised legislative proposals on waste. Key elements of the revised waste proposal include targets for recycling of municipal waste, for recycling of packaging waste, for reduction of municipal waste; economic instruments to discourage landfilling; measures to promote re-use and stimulate industrial symbiosis; economic incentives for producers to put greener products on the market.

The waste-to-bioeconomy transition will deliver increased valorisation of organic waste and the production of bio-derived products. High-value products can be generated from green chemistry, bioprocesses, industrial biotechnology, catalysis and CO2 utilisation. Improved crop breeding and genetic modification could increase crop yield and reduce the use of water, fertilisers and crop protection chemicals. Automation, robotics and precision agriculture approaches could significantly improve resource efficiency in agriculture. Nutrient cycling could reduce dependence on — and the environmental impacts of – fertilizer products. Development and adoption of bio-control measures could reduce reliance on agrochemicals. Better field draining systems or collection systems could reduce run-off of nitrogen and phosphorus. Analysing and sharing data from farms and across the supply chains could identify opportunities to save, share or focus resources. Innovative solutions are required to allow reprocessors and manufacturers to manage the variable nature of waste feedstocks. Developing predictive tools for by-product quality and the stabilization of waste could contribute to this. Improved technologies are required for co-processing packaging and food waste (e.g., compostable/biodegradable packaging).

Centralised de-packaging facilities could allow for a better separation and sorting of packaging materials, which could increase the rate of recovery. Future waste streams such as by-products from the production of protein from insects will generate new valorisation opportunities. There is also still plenty of scope for improved valorisation of existing bio-waste streams such as manure. Developments in biotechnology could allow a wider range of sugars to be suitable feedstocks in biofuel production. Waste gas waste streams from manure and other biological transformations can potentially be valorised. Biomass could be used to clean water and valorise inorganic (e.g. phosphate, nitrate) and organic (sludge) content in waste water [44].

Sustainable Investment

Investors have a critical role to play in financing stronger links across local, regional, and global agriculture value chains.

Responsible investment is an approach to investing that aims to incorporate environmental, social and governance (ESG) factors into investment decisions, to better manage risk and generate sustainable, long-term returns.

Impact Investment is a different framework that can be used by investors when they consider how to invest in sustainable food systems. This framework helps investors to identify high-impact investment opportunities in food, farming, and forestry that align with their specific financial and impact objectives in portfolio management [45].

In the UK, several government departments support research and innovation underlying sustainable food production. Innovate UK and Biotechnology and Biological Sciences Research Council (BBSRC) invested £90 million Sustainable Agriculture and Food Innovation Platform. Innovate UK, BBSRC and the Department for International Development (DFID) jointly invested £70 million in AgriTech Catalyst, a programme funding collaborative projects aiming to develop innovative solutions to global challenges in the agricultural technology. InnovateUK invested £90 million in four Centres for Agricultural Innovation, as part of delivering the UK Strategy for Agriculture Technologies[3].

Conclusion

Sustainability is arguably the single most important global issue. The multiple pressures on the global ecosystem means that the system has an inbuilt lack of resilience. In addition to the anticipated population increases, loss of habitat, climate change and pressure on scarce natural resources, there are a number of unquantifiable factors that are likely to add to the pressures on the system. These include natural disasters, war and civil disturbance and migration pressures. Continued investment in innovation and in new technology will mitigate the threats to the food supply and to the global ecosystem. The key may not be the size or frequency of the innovation but where it acts. As Malcolm Gladwell argues in his book 'The Tipping Point', the world may seem to be immovable and set on a specific course but with the slightest push, in just the right place, it can be tipped. Recognising where that tipping points resides and how to influence it is the most effective way of creating and controlling change in response to the chaos of external factors.

References

[1] Foley, J. A., Ramankutty, N., Brauman, K. A., Cassidy, E. S., Gerber, J. S., et al. (2011). Solutions for a cultivated planet. *Nature*, *478*, 337–342. doi: 10.1038/nature10452

[2] Brundtland, G. H., et al. (1987). *Our Common Future: Report of the World Commission on Environment and Development.* Accessed via http://www.un-documents.net/our-common-future.pdf on 25 August 2017.

---

[3] https://innovateuk.blog.gov.uk/tag/sustainable-agriculture-and-food-innovation-platform/

[3] European Commission (2006). Environmental Impact of Products (EIPRO): Analysis of the Life Cycle Environmental Impacts Related to the Total Final Consumption of the EU 25. *Technical Report EUR 22284 EN.*

[4] *PAS (2050:2011). Specification for the assessment of the life cycle greenhouse gas emissions of goods and services, BSI*, Accessed via https://shop.bsigroup.com/forms/PASs/PAS-2050/ on 25 August 2017.

[5] United Nations (2017). *World population prospects*, 2017 revision. Accessed via https://esa.un.org/unpd/wpp/Publications/Files/WPP 2017_Key Findings.pdf on 23 August 2017.

[6] Ramankutty, N., Evan, A. T., Monfreda, C. & Foley, J. A. (2008). Farming the planet: 1. Geographic distribution of global agricultural lands in the year 2000. *Global Biogeochem Cycles*, *22*, GB1003, doi:10.1029/2007GB002952.

[7] Tilman, D., Balzer, C., Hill, J. & Befort, B. L. (2011). Global food demand and the sustainable intensification of agriculture. *Proc Natl Acad Sci USA*, *108*, 20260–20264. doi: 10.1073/pnas.1116437108.

[8] Lambin, E. F. & Meyfroidt, P. (2011). Global land use change, economic globalization, and the looming land scarcity. *Proc Natl Acad Sci USA*, *108*, 3465–3472. doi: 10.1073/pnas.1100480108.

[9] Kissinger, G., Herold, M. & de Sy, V. (2012). *Drivers of deforestation and forest degradation: a synthesis report for REDD+ policymakers Lexeme Consulting, Vancouver, Canada.* Accessed via https://www.forestcarbonpartnership.org/sites/fcp/files/DriversOfDeforestation.pdf_N_S.pdf on 23 August 2017.

[10] International Union for Conservation of Nature [IUCN]. (2014). *The IUCN Red List of Threatened Species. Version 2014.3.* http://www.iucnredlist.org via the Internet. Accessed 2015 March 14.

[11] Tanentzap, A. J., Lamb, A., Walker, S. & Farmer, A. (2015 Sep). Resolving Conflicts between Agriculture and the Natural Environment. *PLoS Biol.*, *13*(9), e1002242. doi: 10.1371/journal.pbio.1002242

[12] Bondeau, A., Smith, P. C., Zaehle, S., Schaphoff, S., Lucht, W., et al. (2007) Modelling the role of agriculture for the 20th century global terrestrial carbon balance. *Glob Chang Biol*, *13*, 679–706. doi: 10.1111/j.1365-2486.2006.01305.x

[13] Bouwman, L., Goldewijk, K. K., Hoek, K. W. V. D., Beusen, A. H. W., Van Vuuren, D. P., et al. (2013). Exploring global changes in nitrogen and phosphorus cycles in agriculture induced by livestock production over the 1900–2050 period. *Proc Natl Acad Sci USA*, *110*, 20882–20887. doi: 10.1073/pnas.1012878108.

[14] Diaz, R. J. & Rosenberg, R. (2008). Spreading dead zones and consequences for marine ecosystems. *Science*, *321*, 926–929. doi: 10.1126/science.1156401.

[15] Tanentzap, A. J., Lamb, A., Walker, S. & Farmer, A. (2015 Sep). Resolving Conflicts between Agriculture and the Natural Environment. *PLoS Biol.*, *13*(9), e1002242. doi: 10.1371/journal.pbio.1002242.

[16] Greenleaf, S. S. & Kremen, C. (2006). Wild Bees Enhance Honey Bees' Pollination of Hybrid Sunflower. *PNAS*, *103 (37)*, 13890–95. doi: 10.1073/pnas.0600929103.

[17] *United States Environmental Protection Agency (2007). Factors Affecting Pollinator Health.* Accessed via https://www.epa.gov/pollinator-protection/pollinator-health-concerns#factors on 23 August 2017.

[18] Van Beek, L. P., de Graaf, I., Sutanudjaja, E., Wada, Y. & Bierkens, M. F. P. (2006) GC43C-1188: Limits to global groundwater consumption. *Poster presented on 15 December 2016 at the American Geophysical Union conference in San Francisco*, accessed via https://agu confex.com/agu/fm16/meetingapp.cgi/Paper/176019 on 24 August 2017.

[19] Personal communication.

[20] Dalin, C., Wada, Y., Kastner, T. & Puma, M. J. (2017). Groundwater depletion embedded in international food trade. Nature Letter, *Nature*, *543*, pp. 700–704 (30 March) doi:10.1038/ nature21403.

[21] *WRAP (2017). Estimates of Food Surplus and Waste Arisings in the UK.* Accessed via http://www.wrap.org.uk/sites/ files/wrap/ Estimates_%20in_the_UK_Jan17.pdf on 24 August 2017.

[22] *WRAP website.* Accessed on 2/08/2017 via http://www.wrap.org.uk/content/all-sectors

[23] *HM Government (2015). Building a high value bioeconomy - Opportunities from waste.* Accessed via https://www.gov.uk/government/uploads/system/uploads/attachment_data/file/408940/BIS-15-146_Bioeconomy_report_-_opportunities_from_waste.pdf. on 25 August 2017.

[24] *FAO webpage. Agriculture and soil biodiversity* accessed via http://www.fao.org/agriculture/crops/thematic-sitemap/theme/spi/ on 24 August 2017.

[25] *DEFRA website. Soil evidence review tendering note.* Accessed via http://randd.defra.gov.uk/Default.aspx?Menu=Menu&Module=More&Location=None&Completed=1&ProjectID=18786 on 24 August 2017.

[26] *Soil Security Programme website* accessed via https://www.soilsecurity.org/ on 24 August 2017.

[27] *AHDB website. Soils Management webpage.* Accessed via http://www.ahdb.org.uk/projects/soils.aspx on 24 August 2017.

[28] *WRAP (2017). Food futures: from business as usual to business unusual.* Accessed via http://www.wrap.org.uk/sites/files/wrap/Food_Futures_%20report_0.pdf on 24 August 2017.

[29] *Fujitsu (2016). Press release 28 November 2016, Fujitsu Launches Company in Finland to Produce and Sell Vegetables Year-Round with Artificial-Light Plant Factory*, accessed on 02/08/2016 via http://www.fujitsu.com/global/about/resources/news/press-releases/2016/1128-01.html on 24 August 2017.

[30] http://growup.org.uk/

[31] http://www.saturnbioponics.com/ .

[32] http://growing-underground.com/.

[33] Kissinger, G. (2013). *Reducing Risk: Landscape Approaches to Sustainable Sourcing – SABMiller Case Study*, Accessed via

http://peoplefoodandnature.org/wp-content/uploads/2014/06/reducing_risk_sabmiller_case_study.pdf on 25 August 2017 on 25 August 2017.

[34] *WRAP website. What is courtauld.* Accessed via http://www.wrap.org.uk/content/what-is-courtauldand 25 August 2017.

[35] *WRAP website. Courtauld Commitment webpage*. http://www.wrap.org.uk/content/courtauld-commitment-2025-transform-uk-food-and-drink

[36] *M&S website. M&S PlanA 2025*. accessed on 2/08/2017 via https://corporate.marksandspencer.com/documents/plan-a/plan-a-2025-commitments.pdf.

[37] *M&S website. M&S Plan A report*. Accessed via http://planareport.marksandspencer.com/M&S_PlanAReport_2017_FullReport.pdf on 25 August 2017on 25 August 2017 on 25 August 2017.

[38] *WRAP website. Love Food Hate Waste programme: an introduction.* Accessed via http://www.wrap.org.uk/sites/files/wrap/Love%20Food%20Hate%20Waste%20Retailer%20Introduction.pdf on 25. August 2017.

[39] *Love Food Hate waste consumer website*. http://www.lovefoodhatewaste.com

[40] *Love Food Hate partner website.* http://www.wrap.org.uk/content/love-food-hate-waste

[41] *OECD (2011). Industrial Biotechnology and climate change: opportunities and challenges.* Accessed via http://www.oecd.org/health/biotech/49024032.pdf on 25 August 2017.

[42] *Capital Economics (2016). Evidencing the Bioeconomy.* Accessed via https://bbsrc.ukri.org/documents/1607-evidencing-the-bioeconomy-report/ on 25 August 2017.

[43] *European Commission, DG Environment webpage website. Circular Economy.* Accessed via http://ec.europa.eu/environment/circular-economy/index_en.htm on 2 August 2017.

[44] *Knowledge Transfer Network (2016). Innovation Opportunities from Industrial Waste.* Accessed via https://connect.innovateuk.org/documents/2832130/32795723/Innovation+Opportunities+from+Ind

ustrial+Waste/e150620e-4007-4b7c-8730-93f98378739a on 2 August 2018.

[45] Lang K., Humphreys J., and Rodinciuc A. (2017). Impact investing in sustainable food and agriculture across asset classes. Financing Resilient Value Chains through Total Portfolio Activation. Accessed via http://www.trilliuminvest.com/wp-content/uploads/2017/05/Investing-in-Sustainable-Food-and-Agriculture.pdf on 25 August 2017.

In: Food: The Vital Ingredient ISBN: 978-1-53614-402-4
Editors: A. B. Hanley and W. R. Russell

Chapter 7

DIET, HEALTH AND CHRONIC DISEASE

Wendy R. Russell[1] and A. Bryan Hanley[2]
[1] University of Aberdeen Rowett Institute,
Aberdeen, Scotland, UK
[2]The Knowledge Transfer Network, London, UK

INTRODUCTION

There are several definitions of food security, but all agree that an impotent outcome is the provision of 'healthy' life [1]. To maintain human health, it is important that people not only have enough safe food, but when food is available that they are able to make the correct food choices in order to meet nutritional requirements and assuring these options are available for future generations. Considering all aspects of malnutrition; under-, poor and over-nutrition, the majority if the world's population are currently facing substantial risks to health [2]. This is likely to be exacerbated in the face of future challenges to our food supply system through population growth and associated reductions in land mass and fresh water availability to support food production, as well as declining energy supplies, ecosystem preservation and climate change [3, 4]. It is essential that research, policy and practice align globally to tackle these issues. The research challenges are inherently inter-sectorial and each discipline must adjust to new ways of working together. Chemistry has underpinned our understanding of the link between food and human health and will play major role in defining the requirements for future nutritional resilience.

Under-Nutrition, Causes and Consequences

It is estimated that in lower economy countries around one billion people are chronically hungry, meaning that they do not have sufficient food to continuously meet their energy requirements from diet. Wasting, defined by the WHO as below minus two standard deviations from median weight for height of reference population, is a measure of acute malnutrition and is a strong predictor of mortalityIt affects approximately 52 million children (under five) with the highest prevalence in South Asia [5]. For most, this is due to insufficient access to food, though either availability and/or insufficient household income. It is clear that if agricultural practice is to be successful it will have to change from the current paradigm to a more dynamic model (DFID), which will deliver benefits directly to the small-scale growers and local communities. Alternative small-scale farming practices, such as urban and back-yard gardening, as well as commercialisation opportunities for co-operatives should be considered, but of prime importance will be access to crops of higher nutritional quality, which are more efficient with regard to water and energy inputs. Current agriculture is predominantly based on high yielding varieties with favourable harvesting and processability characteristics. These require high-input demands in terms of energy, water, fertilizers, and pesticides. High inputs are needed for maximum production [6], but as these become scarcer and more costly, current crops will not perform as well and this should promote a shift towards developing lower-input varieties [7].

However, in order to maximise benefits (and profitability) varieties should not just deliver improved production performance and processability, but should provide enhanced levels of important nutrients and health promoting constituents. Breeding technologies, including genetic modification and gene editing, accelerated trait identification and marker assisted selection have potential to deliver several nutritional targets and key to realising these benefits will be chemical analysis of both the desired nutrients as well as anti-nutrients including phytates, oxalates, alkaloids, lectins, saponins, tannins cynanogens and chymotrypsin inhibitors [8].

In addition to optimisation of nutrient formation in foods, there is a need to minimise anti-nutritional factors such as those endogenous factors mentioned above. Breeding strategies to reduce levels of alkaloids in lupins was carried out as long ago as 1932 [9] and the greatly increased acreage devoted to rapeseed was in part a consequence of the discovery of varieties such as Bronowski and development of 'double zero' varieties (low glucosinolate, low erucic acid) whose meal after processing to remove the oil had improved nutritional characteristics [10].

Exogenous factors are also an important source of anti- or non-nutritional components in foods. Crops such as rice are able to accumulate heavy metals such as cadmium and arsenic and through targeted breeding strategies and varieties of rice, which accumulate lower levels of these toxins have been identified [11]. Reduction of contamination by specific mycotoxin-producing species (e.g., Fusarium and tricothecenes) through selective breeding is another important target [12]. However, contamination and mycotoxin production is primarily caused by inadequate storage so the effect of breeding strategies to prevent fungal growth is likely to be limited. Chemical technologies could deliver rapid, low-cost diagnostics for the post harvest analysis and quantification of these toxins, specifically for use in LMIC countries and this would be a valuable resource [13]. Several of these detrimental, exogenous compounds can also be reduced by effective food preparation and processing and food chemistry could inform on best practice. It is suggested that detoxification technologies for high-risk commodities, such as mycotoxins should be a priority for research and that possessing biotechnology offers the best potential for future developments

[14]. Current techniques involve primarily separation of contaminated from uncontaminated material at an early stage however some processes (e.g., acid or alkali treatment, fermentation) are able to chemically transform mycotoxins into non-toxic entities. Future developments are likely to include the use of specific detoxification enzyme mixtures. However it should be noted that removal of toxins during processing should not be used as a way of mitigating poor agronomic or storage practices since unknown mycotoxins may also be formed and these may not necessarily be removed during the processing step.

Most of our food both plant and animal-based are affected by soil symbioses, mostly as a consequence of mycorrhizal fungi selectively concentrating and excluding nutrients. In order to produce healthy food, a healthy soil ecosystem is paramount. Identifying solutions to improve soil quality requires robust data to be collected, as predictions suggest that this may have a bigger impact on crop production than climate change [15]. Chemical treatment has an important role to play in effective soil management systems. Understanding the interrelationships between the chemical, physical and biological aspects of the soil, and how to intervene most effectively, is essential to the sustainable use of farming land and to the maximum production of high nutrient crops. In addition, to the soil ecosystem it is estimated that up to 20% of the global population will be exposed to water scarcity at 2 C of global warming [4] and this will impact on the diversity of crops that are able to be produced. Chemistry solutions that could be applied include desalination of water, purification and reuse of waste water as well as developing low-water cropping systems [16].

Where malnutrition results from a lack of access to adequate food, understanding any loss from the food supply chain is essential. Current data on this varies widely and is unreliable and improved efforts for data collection are required [17]. However, it is generally considered that the majority of loss occurs immediately after harvest in developing countries. As developing countries are reliant on small scale farming and short food supply chains, losses are likely to be mostly due to inadequate processing and storage technology [18]. This is exacerbated by agricultural productivity which, in Sub-Saharan Africa, for example, averaged 2.3 per cent per year in the 1980s and 3.8 per cent during the 2000s. Not only is this less than population growth, it has largely been based upon expansion of agricultural land areas rather than increases in productivity [19]. A combination of improved agronomic practice and the development of food processing without concomitant high infrastructure costs is a clear requirement. A recent report outlined some of the key areas where prevention of food loss could be made [20]. The authors pointed out that losses of fruit and vegetables globally leads to the loss of equivalent iron content for up to 66 million people and vitamin A content for up to 70million people. Factors such as agronomic practice, better packaging and access to a cold chain are key.

In addition to chronic hunger, another two billion individuals suffer from micronutrient deficiency [21]. This includes both vitamins and minerals and demonstrates a problem beyond energy and macronutrient availability. This wider-scale problem is due predominantly to consumption of diets of low nutritional quality and those lacking diversity. Again, poverty and access to a wide range of dietary sources are predominantly responsible, but in higher economy countries, poor dietary choices are a contributory factor. In particular, according to the National Diet and Nutrition Survey, females age 11-18 are more than 20% below the lower reference nutrient intake for manganese, iron, selenium, potassium, iodine, riboflavin, calcium and zinc, with iron, potassium and selenium remaining below this level for female age 19-64 [22]. Globally, iron, vitamin A, iodine and folate are the most common deficiencies, with zinc and vitamin D becoming an increasing concern [5, 23].

Chronic deficiency in micronutrients can have long-term consequences. In lower economy countries, stunting results in both cognitive and physical retardation and although rates have declined around 200 million children remain severally stunted [24]. Micronutrient deficiency does not just impact on lower-economy countries. In the EU it is estimated that 50 million people face deprivation, with 18 million receiving food aid [25]. As micronutrient deficiencies are widespread across different demographics, it is essential that a better understanding of individual requirement is achieved. Whether this is delivered addressed through improvements in soil structure and crop production, or through fortification and reformulation strategies will be circumstantial and must meet social acceptance.

OVER-NUTRITION

Worryingly, more than one third of all adults are overweight or obese and this is increasing worldwide [26]. This has an enormous cost to both the economy and public health, but also to our global food supply system. Over-nutrition does not necessarily equate with an adequate intake of nutrients and indeed is mostly inversely associated with health. Lifestyle associated diseases, some of which reflect inadequate nutrient intake, such as cardiovascular disease, type 2 diabetes mellitus and cancer are associated with over nutrition [27, 28]. The world's urban population is now greater than those living rurally. According to the UN, in 2014, the urban population comprised 54% and this was expected to increase to 66% by 2050. This means that the urban population will be greater than 6 billion by 2050. The dislocation between food production and consumption will continue to grow. Lifestyle-associated disorders associated with over-nutrition are now found alongside under-nutrition, as citizens from emerging economies become prosperous and dietary patterns shift (e.g., greater animal protein consumption) and concomitantly impact on fresh water and energy use. One of our major challenges in delivering food security will be to encourage sustainable eating patterns and reducing food waste, as countries continue to increase in economic status and to try to reverse the trend towards over-consumption in developed countries.

In addition to micronutrient and nutrient intake, a number of other dietary components contribute to a healthy lifestyle – particularly in the context of chronic health. Many of these could underpin health and wellness by preventing or delaying the onset of chronic disease. It will only be possible to encourage consumers to choose healthier options, if the research supporting the concept of healthy foods is robust. The strongest evidence for efficacy is obtained through conducting well designed randomised controlled trials (RCT) in humans with established biomarkers and well-defined foods. The European Food Safety Authority (EFSA) will not recognise a health claim if the food product has not been adequately characterised in terms of its claim target constituents (chemical or biological). So in addition to having validated biomarkers and a clear dose: effect relationship, this must all be underpinned by clear evidence of which food component is responsible for the beneficial effect.

One of the constituents for which there is currently strong evidence of beneficial effects on health is fibre; both soluble and insoluble. In addition to improving laxation and regularity of bowel movement [29], epidemiological evidence suggest a strong association with fibre and cardiovascular health and that adequate intake lowers the risk of both cardiovascular disease and coronary heart disease [30]. The evidence for randomised clinical trials is also increasing and soluble fibre such as β-glucan and psyllium have been shown to lower serum LDL cholesterol concentrations and health claims have been approved in both Europe and the US [31, 32]. Observational evidence also suggests that regular fibre consumption of more than 15 g/day significantly reduces the risk of type II diabetes [33]. Intervention for nutritional control of blood glucose also showed the strongest evidence for both soluble and insoluble fibre [34]. However, carbohydrate constituents of the diet are generally poorly characterised and often rely on measurement of the individual sugars from which they are comprised following hydrolysis, rather than structural analysis, which would provide important information on their mechanism of action.

There is also substantial evidence that phytochemical-rich foods; fruits, vegetables, grains and cereals are beneficial. But this evidence is weak, due to poor characterisation of the active components and a lack of understanding of their metabolism and of which of these metabolites might constitute the active principle(s). Plant-based foods contain several thousands of bioactive phytochemicals and research to characterise these may lead to identifying a functional role for these potentially undiscovered vitamins.' Although protein and fats are generally better characterised, for all dietary constituents there is a lack of knowledge regarding their metabolism in vivo. As increasing evidence suggests a major role of the gut microbiota in the development of dietary-related diseases, it is essential that an understanding of the metabolic fate of food components is established. Never has it been timelier, that food technology can deliver knowledge that will allow an understanding of the relationship between diet and disease. It will facilitate the discovery of effective biomarkers, keystone bacteria and more importantly strategies to tackle the increasing burden of these diseases to human health, the economy and delivering a secure food system.

As well as encouraging consumers to make the correct food choices, engagement across the food supply will ensure provision of healthy options. If the important constituents of food are identified, crop selection and breeding strategies can be adopted to deliver both healthy and sustainable ingredients. From an agricultural perspective, reduction of fertilisers and pesticides is a much needed area of research. Plant breeding, perhaps considering ancient and wild varieties could be useful in reintroducing some of the bioactive components that may have bred out of our food through domestication,

As well as having healthy and sustainable food products available, it will be essential that these ingredients are developed into processed foods. Food reformulation is an ideal opportunity to improve health and needs to go beyond reduction of energy, fat, salt and sugar (see Chapter 8). From a health perspective, reformulation should consider increasing fibre and phytochemical-rich ingredients, as increasing evidence supports their importance in the diet Ingredients should also be sourced from a sustainability standpoint. In particular, identifying sustainable sources of protein is key. These may come from animal sources such as insects or through sustainable intensification of agricultural practice. Micro-organisms might also prove useful and can be used to create protein from simple carbon sources. However, when considering both health and sustainability, high-protein plants have potential to deliver both significant amounts of protein, as well as additional benefits from fibre, micronutrients and phytochemicals.

For example, while the protein content of rice is relatively low, it can be processed to give yields of the protein glutelin by alkaline extraction. Indeed rice bran protein (a by-product of rice milling) has been suggested to be a useful protein source if processed using a range of non-alkaline techniques [35]. Rice starch can be subjected to retrogradation to increase the fibre content. The Central Rice Research Institute in India has developed a high protein rice (protein content around 11.5% compared to 6% or so in commonly cultivated varieties) by crossing a protein rich germplasm with high yielding varieties. The intention is to further enhance the nutritional value of rice by creating varieties with high zinc and iron content. If ingredient selection is addressed across the food supply chain, reformulation can provide opportunities for growers both at a national and global level.

CONCLUSION

Chronic disease is both a contributor to and a consequence of the current agri food paradigm. As populations grow then there will be a perfect storm of effects on increasing food production to deal with increased populations and enhanced expectations, improving the nutritional efficiency of the whole food chain and dealing with the increasing burden of chronic disease caused both by under and over nutrition but also a consequence of increased life expectancy. Priorities will shift from maximising yield and profit to improving diet and health in a more holistic way – taking into account different nutritional needs at different life stages and changes in outlook and perception while at the same time maintaining the production of food that is desirable, nutritious and healthy.

Delivering healthy and sustainable food security involves the chemical sciences working not just across the food supply chain, but closely with several other disciplines. It is only through truly transdisciplinary research and stakeholder engagement will realising these objectives be possible (Table 1).

Table 1. Examples whereby multidisciplinary research can improve the diet and health aspects of food security

Opportunity	Disciplines
Improving soil health through understanding optimum requirements and methods of obtaining these levels	soil scientists agronomists
Plant breeding strategies to both minimise toxins and anti-nutrients, as well as increase nutrients vital nutrients	plant breeders toxicologists
Development of low-water agricultural systems, including efficient desalination	engineers agronomists
Ensuring food safety through low cost robust diagnostics, in particular for post-harvest analysis transport and storage	technologists toxicologists
Minimising food loss though novel monitoring systems and improved methods of distribution	technologists engineers
Understanding nutritional requirements for particular demographics, consideration of personalised nutrition	nutrition scientists dieticians
Development of biofortification and fortification strategies	nutrition scientists dieticians food formulation scientists
Delivering robust evidence for defining healthy food to encourage dietary change, understanding human metabolism, biomarker discovery	nutrition scientists human biologists social scientists
Identifying detrimental consequences of domestication and current agricultural practice, identifying sustainable solutions	agronomists plant scientists
Reformulation of current foods beyond salt, sugar, fat and energy to include dietary components for which there are strongest evidence for health benefit	food formulation scientists nutrition scientists

Consideration of both down- and up-stream targets of research programmes can add value to this process. An exemplar of this could be the reformulation of foods to reduce chronic disease in higher economy countries, delivering opportunities for sustainable ingredient production in low and middle income countries combines with the development of small scale, rapid processing in producer countries. Researchers working in chemistry must develop these connections, but research governance and funding bodies must play a greater role in supporting these joined up initiatives and ensuring that sufficient emphasis is given to the social as well as the economic factors that influence and impact upon people at a population and individual level. While much can be achieved through scientific endeavour, there is also an important role for legislation and appropriate safeguards to be enshrined in law but this will require a more coherent policy framework and integrated response [36].

REFERENCES

[1] *FAO 2009 State of food insecurity.* 2009 Food and Agriculture Organization, Rome.

[2] Gardner, G. and Halweil, B. Overfed and Underfed: The Global Epidemic of Malnutrition (2000). *Worldwatch Paper.* Jane Peterson Eds. Washington, DC.

[3] Freibauer A., Mathijs E., Brunori G., Damianova Z., Faroult E., Girona, I., Gomis J., O'Brien L. and Treyer S. *2011 Sustainable food consumption and production in a resource-constrained world: the 3rd SCAR Foresight Exercise European Commission Standing Committee.*

[4] Schewe, J., Heinke, J., Gerten, D., Haddeland, I., Arnell, N.W., Clark, D. B., Dankers, R., Eisner, S., Fekete, B. M., Colón-González, F. J., Gosling, S. N., Kim, H., Liu, X., Masaki, Y., F. T. Portmann, Satoh, Y., Stacke, T., Tang, Q., Wada, Y., Wisser, D., Albrecht, T., Frieler, K., Piontek, F., Warszawski, L., and Kabat, P. Multimodel assessment of water scarcity under climate change (2014) *PNAS,* 111/ :3245–3250.

[5] Black, R. E., Allen, L. H., Bhutta, Z. A., Caulfi eld, L. E., de Onis, M., Ezzati, M., et al. (2008) Maternal and child undernutrition: global and regional exposures and health consequences. *Lancet,* 371, 243 – 260.

[6] Aleklett, K., Hook, M., Jakobsson, K., Lardelli, M., Snowden, S., Soderbergh, B. The peak of the oil age—Analyzing the world oil production reference scenario in world energy outlook 2008. *Energy Policy* 2010, 38, 1398–1414.

[7] Fess, T. L., Kotcon, J. B., Benedito, V. A. *Crop Breeding for Low Input Agriculture: A Sustainable Response to Feed a Growing World Population Sustainability* 2011.

[8] Bevan M. W., Uauy C., Wulff B.B., Zhou J., Krasileva K., Clark M. D. Genomic innovation for crop improvement. *Nature.* 2017; 543(7645):346-354.

[9] Ivanov, N. N., Smirnova, M. I., Sharapov, N. I., Fedotov, V. S. et al. 1932. The Problem of alkaloidness in lupin. *The Exibit No 54 to Bull. Apl. Bot. Gen. Pl.-Breed.,* Leningrad-Moscow, 1932, 63 p.

[10] Downey, R. K., 1990: Brassica oilseed breeding-achievements and opportunities. *Plant Breeding Abstracts*, 60 (10): 1165-1170.

[11] Guilan Duan, Guosheng Shao, Zhong Tang, Hongping Chen, Boxun Wang, Zhu Tang, Yuping Yang, Yuechuan Liu and Fang-Jie ZhaoGenotypic and Environmental Variations in *Grain Cadmium and Arsenic Concentrations Among a Panel of High Yielding Rice Cultivars* (2017) Rice 10:9.

[12] Buerstmayr, H. and Lemmens, M. Breeding healthy cereals: genetic improvement of Fusarium resistance and consequences for mycotoxins. *World Mycotoxin Journal*: 8 (5) - Pages: 591 – 602.

[13] Alex, P. Wacoo, Deborah Wendiro, Peter, C. Vuzi, and Joseph, F. Hawumba Methods for Detection of Aflatoxins in *Agricultural Food Crops Journal of Chemical Technology & Biotechnology* 2014.

[14] Petr Karlovsky, Michele Suman, Franz Berthiller, Johan De Meester, Gerhard Eisenbrand, Irène Perrin, Isabelle P. Oswald, Gerrit Speijers, Alessandro Chiodini, Tobias Recker and Pierre Dussort Impact of food processing and detoxification treatments on mycotoxin contamination *Mycotoxin Research* 2016, 32/4, pp 179–205.

[15] Christian Folberth, Rastislav Skalský, Elena Moltchanova, Juraj Balkovič, Ligia B. Azevedo, Michael Obersteiner & Marijn van der Velde Uncertainty in soil data can outweigh climate impact signals in global crop yield simulations *Nature Communications* 7, 11872 (2016).

[16] Lienhard, J. H., Mistry, V. K. H., Sharqawy, M. H. and Thiel, G. P. "Thermodynamics, Exergy and Energy Efficiency in Desalination Systems," in *Desalination Sustainability: A Technical, Socioeconomic, and Environmental Approach, Chpt. 4*, H.A. Arafat, ed. Elsevier Publishing Co., 23 June 2017.

[17] Parfitt, J., M. Barthel and S. Macnaughton (2010) "Food waste within food supply chains: quantification and potential for change to 2050." *Philosophical Transactions of the Royal Society – Biological Sciences*: 3065-3081, UK.

[18] Jayne T. S., Zulu B., Nijhoff J. J. 2006 Stabilizing food markets in eastern and southern Africa. *Food Policy* 31, 328–341.

[19] Nicolas Depetris Chauvin, Francis Mulangu and Guido Porto, Food *Production and Consumption Trends in Sub-Saharan Africa: Prospects for the Transformation of the Agricultural Sector*, United Nations Development Programme, WP 2012-011: February 2012.

[20] Winkworth-Smith CG, Foster, TJ and Morgan W (2015) *The impact of reducing Foos Loss in the Global Cold Chain* http://naturalleader.com/wp-content/uploads/2016/04/UTC-Nottingham-Report_3-30_FINAL.pdf.

[21] *The World Health Report World Health Organization 2000.*

[22] Bates B, Lennox A, Prentice A, et al. (2012) *National Diet and Nutrition Survey: Headline Results from Years 1, 2 and 3 (Combined) of the Rolling Programme 2008/09–2010/11.* Chapter 5 Tables. Department of Health. London: The Stationery Office.

[23] Shetty, P. and Schmidhuber, J. Nutrition, lifestyle, obesity and chronic disease (2011). *Expert Paper 2011/3.* United Nations Department of Economic and Social Affairs.

[24] *UNICEF 2013.*

[25] The role of research in global food and nutrition research. *Expo 2015 EU Scientific Steering Committee.*

[26] NCD Risk Factor Collaboration (NCD-RisC).Worldwide trends in body-mass index, underweight, overweight, and obesity from 1975 to 2016: a pooled analysis of 2416 population-based measurement

studies in 128·9 million children, adolescents, and adults. (2017) *Lancet* 16;390(10113):2627-2642.

[27] Lavie CJ, Arena R, Alpert MA, Milani RV Ventura HO (2018) Management of cardiovascular diseases in patients with obesity. *Nature Reviews Cardiology* volume 15, pages 45–56.

[28] Al-Goblan AS, Al-Alfi MA and Khan MZ Mechanism linking diabetes mellitus and obesity *Diabetes Metab Syndr Obes*. 2014; 7: 587–591.

[29] Cummings J. H. The Effect of Dietary Fiber on Fecal Weight and Composition. In: Spiller G. A., editor. CRC *Handbook of Dietary Fiber in Human Nutrition.* CRC Press; Boca Raton, FL, US: 1993. pp. 263–333.

[30] Slavin J. L. Position of the American Dietetic Association: Health implications of dietary fiber. *J. Am. Diet. Assoc*. 2008;108:1716–1731.

[31] Adam M. Bernstein, Brigid Titgemeier, Kristin Kirkpatrick, Mladen Golubic, Michael F. RoizenMajor Cereal Grain Fibers and Psyllium in *Relation to Cardiovascular Health Nutrients*. 2013 May; 5(5): 1471–1487.

[32] Kellen V. Lambeau, Johnson W. McRorie, Jr. Fiber supplements and clinically proven health benefits: How to recognize and recommend an effective fiber therapy *J Am Assoc Nurse Pract*. 2017 Apr; 29(4): 216–223.

[33] Hopping BN, Erber E, Grandinetti A, Verheus M, Kolonel LN, Maskarinec G Dietary fiber, magnesium, and glycemic load alter risk of type 2 diabetes in a multiethnic cohort in Hawaii. *J Nutr*. 2010 Jan; 140(1):68-74.

[34] Russell WR, Baka A, Björck I, Delzenne N, Gao D, Griffiths HR, Juvonen K, Lahtinen S, Lan-sink M, van Loon L, Östman E, Riccardi G, Vinoy S and Weickert MO (2013) Impact of diet com-position on blood glucose regulation. *Critical Reviews in Food Science* 56(4):541-590.

[35] Phongthai S, Homthawornchoo W and Rawdkuen S (2017) Preparation, properties and application of rice bran protein: *A review Int Food Research Journal*, 24, 25, 2017).

[36] Laing T. and Barling D. Food security and food sustainability: reformulating the debate 1 *The Geographical Journal*, Vol. 178, No. 4, December 2012, pp. 313–326 on Agricultural Research (SCAR), Brussels.

In: Food: The Vital Ingredient ISBN: 978-1-53614-402-4
Editors: A. B. Hanley and W. R. Russell

Chapter 8

FOOD PROCESSING MANUFACTURING AND REFORMULATION PROCESSES

Madalina Neacsu**, PhD
The Rowett Institute, University of Aberdeen,
Aberdeen, UK

INTRODUCTION

The Food and Agriculture Organization (FAO) identified four dimensions to food security: food availability, economic and physical access to food, food utilization and stability over time [1]. Food manufacturing, processing and reformulation are key elements to ensuring the food supply is safe, can be accessed in a suitable form and is appealing to consumers.

---

* Corresponding Author: m.neacsu@abdn.ac.uk.

FOOD PROCESSING FOR SAFETY AND QUALITY

Fruit and Vegetables

Minimal processing is essential for fresh fruit and vegetables. These require proper handling, preparation and storage and this includes unit operations such as washing, trimming, peeling, cutting and shredding, sanitizing and packing. Ensuring food safety and maintaining nutritional and sensory quality of preservation minimally processed fruits and vegetables, require refrigeration at 4°C [2-4], which extends the product shelf-life to 5 to 7 days [5].

There are many types of microorganisms that can be found on fresh fruit and vegetables and their number is highly variable. *Pseudomonas spp.* dominate and may make up 50–90% of the microbial population on many vegetables [6, 7, 8]. Contamination of fresh fruits and vegetables is of special concern, because such produce is likely to be consumed raw, and safe production methods and proper disinfection/decontamination procedures are therefore critical steps in ensuring the safety of fresh fruits and vegetables [9-12]. In addition, good hygiene is essential to ensure raw product safety. The US Center for Disease Control suggested that up to 46% of food poisoning cases in the US might be attributed to fruit and vegetables and nuts [13]. The responsible agent in many cases was Norovirus which is associated with poor hygiene. There are a variety of chemical and physical methods used to reduce microorganism population on whole and fresh-cut produce and proved to be efficient in reducing the indigenous microflora and the contaminating pathogens [14, 15]. Oxidation is the second most important cause of food deterioration after that induced by microbiological contamination. The main oxidative reactions are enzymatic browning caused bypolyphenoloxidase and peroxidase [16]. The dark spots on fruits such as bananas, on the other hand, can be attributed to the presence of melanin. Whatever the enzymic basis for browning, its prevention is important to appearance and to consumer acceptability of the product. A range of treatments are used to combat microbiological spoilage and browning.

Antimicrobial Chemical Processes

Typical anti-microbial treatments include the use of chlorine and related bleaches and hydrogen peroxide. Organic acids are also used (e.g., lactic, ascorbic). These chemical treatments are mainly for antimicrobial and anti-browning usage. Other chemical treatments include the use of nitric oxide, sulphur dioxide, ozone, ethylene, 1-methylcyclopropene and controlled atmospheric storage. As an emerging technology, plasma has been successfully used for inactivation of microbial spores [15].

Chlorine (Hypochlorite)

Chlorine solutions are often used to rinse fresh and freshly-cut fruit and vegetables with concentrations varying from 50 to 200 ppm and with typical contact times of less than 5 minutes [17, 18]. Chlorine rinses have shown to decrease the bacterial load by values ranging from 1 to 3.15 log CFU/g, depending on inoculation method, chlorine concentration, contact time, and the target bacteria [19-22].

Chlorine Dioxide has two and the half times greater oxidant capacity compared to chlorine and does not participate in chlorination reactions that result in harmful by-products [23]. It is highly effective against pathogenic organisms such as *Legionella, Giardia cysts, Escherichia coli, Cryptosporidium* and Viruses [18, 24-26]. Efficiency of use depends on concentration, time of exposure, relative humidity, and temperature [27-31].

Acidified Sodium Chlorite

Applications of acidified sodium chlorite performed on raw fruits and vegetables as a spray or dip in the range of 500 to 1200 ppm showed a strong antimicrobial effect against *E. coli* O157:H7 and *Salmonella*. There is a need for more information on the use of acidified sodium chlorite in produce sanitizing [9, 10, 14, 32].

Bromine and Iodine

When added to chlorine solutions, bromine has a synergistic antimicrobial relationship. However, there are safety concerns about the production of brominated organic compounds and their impact on human and environmental safety [14, 32]. Elemental iodine and non-ionic surfactants are less corrosive than chlorine at low temperatures and have a broad spectrum of antimicrobial activity. However, due to the reaction between iodine and starch resulting in a blue-purple colour, the use of iodine containing solutions as sanitizers is limited [4, 9, 33].

Quaternary Ammonium Compounds (QACs)

QACs are cationic surfactants that are able to penetrate food contact surfaces more readily than other sanitizers [9]. They have a strong antimicrobial activity against the fungi and gram-positive bacteria and less strong activity against gram-negative bacteria [34, 35].

Organic Acids

Lactic, citric, acetic or tartaric acid are strong antimicrobial agents due to environmental and intercellular pH reduction, disturbance of membrane transport and/or permeability and anion accumulation [14]. Citric and ascorbic acids are commonly used in fruit and vegetable washing and added to fruit juices [9, 36].

Hydrogen Peroxide

Due to its strong oxidizing power and generation of hydroxyl radical, hydrogen peroxide possesses bacteriostatic and bactericidal activity [9]. For antibacterial use, an optimum range of 2–4% concentration is recommended, since higher concentrations would interfere with the quality of the produce and a lower concentration would be inefficient [18, 20, 37]. It was successfully used in washing waters as an antimicrobial and bleaching agent in the range of 0.04–1.25% up to 80 ppm [38-40].

Peroxyacetic Acid

Peroxyacetic acid is commercialized as a liquid and is a combination of peracetic acid and hydrogen peroxide [9]. Due to its strong oxidant capacity, it is used to wash fruits and vegetables in concentrations up to 80 ppm. However this concentration has shown to be ineffective for substantial reduction in the microbial load of fresh-cut fruits and vegetables [10, 18,

41]. Peroxyacetic acid is effective on the inactivation of pathogenic microorganisms in suspension at lower concentrations than the ones required when using chlorine [9].

Calcium-Based Solutions

Calcium lactate is one of the most commonly used antibacterial due to its ability to uncouple microbial transport processes. In general, calcium solutions are widely used for delicate fruits and products as it maintains produce firmness [9]. Similar effectiveness to chlorine has been demonstrated in reducing microbial load in fresh-cut vegetables [18, 42].

Ozone

Ozone is a powerful antimicrobial agent, which has a high reactivity, penetrability and spontaneously decomposes to a non-toxic product [43, 44]. Ozonated water has successfully extended the shelf-life of fresh-cut fruit and vegetables by reducing microbial load [20, 45]. Treatment with ozone has been shown to be beneficial in extending the storage life of fresh various uncut fruit and vegetables by reducing the microbial load and by oxidation of ethylene [32, 43, 46]. Ozone has a greater effect against certain microorganisms than chlorine and rapidly decomposes to oxygen [47]. In contact with organic matter, ozone creates aldehydes, ketones and carboxylic acids, causing less regulatory concern than chlorine [48]. The main disadvantages of ozone use are the higher corrosive properties and initial capital cost for generator facilities in comparison with the use of chlorine [49].

Electrolysed Water

Electrolysed water is generated by the electrolysis of aqueous salt solutions [50]. Acidic electrolysed water (pH 2.1-4.5) has a strong bactericidal effect against pathogens and spoilage microorganisms, more effective than chlorine due to a high oxidation reduction potential [51] and has shown higher effectiveness in reducing viable aerobes than ozone [52]. Electrolysed water inactivates *Staphylococcus aureus, Pseudomonas aeruginosa, Enterococcus sp., Aeromonas sp. Escherichia coli* and

Legionella pneumophila, also microorganisms such as *Bacillus subtilis, Ralstonia solanacearum, Rhizoctonia solani, Pythium ultimum, Fusarium oxysporum f.sp. lycopersici* and *Trichoderma hamatum* [53]. However, compared to its use in hospitals and healthcare facilities, electrolysed water is less common in food systems due to issues with corrosion and effects on the quality of the treated products.

Natural Antimicrobials

As a response to occasional intolerances in sensitive individuals and the formation of potentially harmful by-products (e.g., nitrosamines from nitrite) of traditional food preservatives, the interest in natural antimicrobial compounds and their consumer demand has increased [54, 55]. Preliminary studies reported by Martin-Diana et al., 2006, [56] with whey permeate in fresh-cut products showed a good antimicrobial activity. The combination of natural antimicrobial with a synthetic chemical were studied; pediocin and nisin applications in combinations with organic acids caused a significant reduction of native microflora and inoculated populations on fresh produce [54, 57].

Chemical Treatments to Prevent Browning

The chemical treatments to limit the oxidation phenomenon of the fruit differ by their action, depending on the used chemical agents: antioxidant agent, chelating agent, firmness agent and acidifying agent [16]. In practice, various combinations of chemical treatments are used, depending on the type of fruit, which needs treatment since different fruits have a varying sensitivity to oxidation due to their structure and composition (Table 1).

Physical Methods for Reducing the Microbiological Load

Some physical methods are also available for reducing the microbiological load and to improve the fruit and vegetables quality and safety. These include modified atmosphere packaging, active and intelligent packaging, nanocomposite packaging, irradiation, ultraviolet light, pulsed light, high pressure processing, ultrasound, cold plasma, and combined methods [9].

Table 1. Chemical treatment for preventing the browning of fruits and vegetables

Treatment	Mode of Action	Examples
Antioxidants	Reaction with oxygen	Ascorbic acid Hexylresorcinol Cysteine
Chelating Agents	Removal of metal ions required for enzyme activity	Citric acid EDTA Kojic acid
Firmness Agents	Strengthens the cell wall and maintains compartmentalisation of enzymes	Calcium salts
Acidifying Agents	Inhibits polyphenol oxidase activity	Citric acid Ascorbic acid

Anti-Browning Methods

There are various physical treatments with different actions like modification of the temperature of the product or a decrease of the availability of oxygen availability of oxygen which contribute to prevention of browning. In blanching and freezing methods, temperature plays a key role since polyphenoloxidase is sensitive to temperature variations, especially to high temperatures (above 80°C) [58]; whereas freezing induces a decrease of available water for the enzymatic reactions leading to less activity of polyphenol oxidase [59]. Another physical treatment used in anti-browning is the conservation under modified atmosphere, which reduces oxygen content and avoids the reaction of enzymatic browning [60].

Soft Drinks

Soft drinks can be classified on the basis of their sugar and fruit juice content, flavouring, carbonation level, main non-water ingredients, and functionality [61]. According to Canadean Beverage Categories Definitions, the most popular soft drinks are as follows: ready-to-drink essence-flavoured beverages; ready-to-drink beverages containing fruits or fruit juice; beverages ready-to-drink after dilution [62]. Functional drinks include drinks enriched with juices, vitamins, and minerals; sports and energy drinks; wellness drinks and nutraceuticals [61]. The ingredients which are commonly identified in the soft drinks are: water, sweetener (8 ÷ 12%, w/v), sugar (1 ÷ 12%, w/w), carbon dioxide (0.3 ÷ 0.6% w/v), acidulants (0.05 ÷ 0.3% w/v), flavourings (0.1 ÷ 0.5% w/v), colourings (0 ÷ 70 ppm), chemical preservatives (lawful limits), antioxidants (<100 ppm), and/or foaming agents (e.g., saponins up to 200mg/mL) [61].

Chemical Preservation

The types of chemical preservatives used to improve the microbiological stability of the drink, depend on the chemical and physical properties of both the preservative and the beverage. The preservative should be used to prevent microbial growth and is dictated by the pH of the product, the presence of vitamins, the packaging, and the conditions of storage [61]. Most used antimicrobial chemicals are: sorbates, benzoates, dimethyl dicarbonate, sulfites and natural compounds (Table 2).

Table 2. Antimicrobial chemicals for preservation of soft drinks

Antimicrobial	**Characteristics**	**Use**
Sorbates and benzoates	Wide-scale activity against yeast, mould and bacteria especially at low pH	Carbonated, non-alcoholic and juice beverages
Dimethyl dicarbonate	Active against yeast, mould and bacteria. Breaks down in water	Cold sterilised soft drinks
Sulfites	Enhance flavour and freshness. Associated with allergic reactions and subject to labelling.	Fruit juices, wine.
Natural Compounds	Various characteristics. May also be a flavour and regulatory penalty	Limited

Packaging

Packaging is a key component for preservation and for maintaining safety and quality. The most commonly used material in the food industry are polyethylene terephthalate for soft drink bottles and water bottles; low density polyethylene for wrapping films and grocery bags; polypropylene for syrup bottles, yogurt tubs, and bottle caps; polystyrene for single-use coffee cups. Soft drinks are also packaged in tin-free steel, aluminium, treated cardboard cartons, or foil pouches [61].

Meat and Meat Products

Meat and meat products are prone to microbial spoilage during slaughtering, processing and storage because they possess an ideal nutrient matrix that can favour the proliferation of micro-organisms which can affect both safety and quality [63]. Meat and meat product contamination could be of bacterial nature; the most commonly reported genera of bacteria on meat surface include *Sheela spp., E. coli, B. protest, Salmonella spp., S. epidermidis, S. aureus and B. cereus* [65]. Yeast could also be a meat contaminant and the yeast species that can significantly affect meat and hence cause spoilage include *Candida mesenterica, Candida saitoana, Cryptococcus albidus, Cryptococcus laurentii, Cryptococcus luteolus, Rhodotrula glutinous, and Rhodotorula mucilaginosa* [66]. Meat spoilage could be of mould nature, the most common fungal species of meat spoilage interest include *Acremonium, Alternaria, Aspergillus, Cladosporium, Epicoccum,* and *Penicillium* [67].

Chemical Technologies

The most common used chemical technologies to keep meat safe for consumption are shown in Table 3: the use of organic acids, chlorine, acidic calcium sulfate, carbon dioxide, activated lactoferrin, peroxycacetic acid, cetylpyridium chloride, ozone, and essential oils and plant extracts [63].

Table 3. Chemical technologies to keep meat safe for consumption

Treatment	Mode of action	Use
Organic acids	Lowering pH	Surface of meats
Trisodium phosphate	Increase pH	Meat tissue
Chlorine	Oxidative potential	Not allowed in the EU but permitted in the US
Peroxyacetic acid	Oxidative potential	Cold and hot meat carcasses
Acidic calcium sulfate	Bacteriostatic effect	Post processing dipping solution
Lactoferrin	Cellular disruptive agent and iron binding	Meat and poultry
Cetyl pyridinium chloride	Bacterial respiration	Surface cleaning

Physical Techniques

Apart from chemical technologies to control microbial spoilage of meat and meat products, there are the *biological intervention technologies* and the *physical technologies*. The biological technologies, include the use of bacteriophages and bacteriocins, which have shown good potential to decontaminate microbes and are used in meat and food industries [63, 67, 68]. Food safety status of meat products can also be improved by deploying natural ingredients like several plant extracts and their essential oils to hinder the growth of microflora [69]. The physical processing technologies that are used for bacterial reduction in meat products include: ionizing irradiation, hydrostatic pressure, electric fields, pulsed light, various packaging technologies, such as vacuum, modified atmosphere, active, antimicrobial; sonication and microwaves [70].

The gaps and challenges currently faced by the food processors translate into the necessity to meet the food sustainability aspects. These mean: security of quality supply chain- a global model adapted locally to meet the food needs of the population; the global harmonisation of regulatory aspects in order to keep the food safe; and to face the consumers demand by providing economic, nutritious and healthy products using energy efficient and environmental friendly technologies.

FORMULATION AND REFORMULATION OF FOOD FOR HEALTH AND NUTRITION

There are a number of ways in which manufacturers, retailers and the food service sector can work with others to improve diet and health. Changing the composition of food (reformulation) is one of the categories of action, which has already been adopted. Within this category, the following main aspects of reformulation were identified as having a health benefit: reducing energy density (kJ/g food); reducing the amount of fat, saturates, sugar or salt; changing the fat or carbohydrate profile [71].

Reduction of Salt, Sugar and Fat

Salt

Salt reduction is one of the primary goals in prevention and control of non-communicable diseases blood pressure in salt sensitive members of the population [72]. There are several national and international initiatives and action plans to reduce the salt intake from the diets in order to meet WHO recommendations of an intake of less than 5 g/day of salt (< 2 g/day sodium) for adults [73, 74]. Across European countries, Eastern and Southern countries have the highest consumption on salt with Czech Republic, 13.6 g/day and the lowest consumption being reported on Germany at 6.3 g/day [75]. It has been estimated that about 75–80% of dietary salt is obtained through processed food consumption, 5–10% is naturally occurring in the foods that make up the diet and the remaining 10–15% comes from salt added during cooking or at the table [74, 75, 77].

In most European countries, a small number of food items are responsible for the highest proportion of salt intake, i.e., bread, cereals and bakery products followed by meat and meat products, then cheese and dairy products [75, 78].

Salt in Baked Foods

The most important function of salt in bread is the maintenance/stabilization of the gluten structure. Salt also inhibits the growth of yeast, controlling the fermentation rate [79]; contributes to flavour and reduces spoilage by controlling water activity in baked products [80]. Salt reduction can result in excessive fermentation, therefore it is important to adapt the proofing time [81]. With increasing yeast activity the amount of free reducing sugars available for Maillard reactions decline and food can lose both visual appeal and taste characteristics. Potassium chloride can be used in bread and other baked products, as it strengthens gluten structure [82], but can only be used up to 10% replacement of sodium chloride, since it negatively affects the taste [80]. Another possibility is the use of magnesium salts. This has been successfully applied with over 30% of sodium chloride being substituted with a mixture of magnesium chloride, potassium chloride and magnesium sulphate [83]. In cakes, sodium chloride plays a minor technological role and is primarily added to improve flavour. Whey permeate, was also successfully used to produce low sodium bakery products [84].

Salt in Meat/Processed Meat

The highest levels of salt was observed in cured meat products and sausages [85]. Salt increases the hydration and water binding capacity of proteins, improves the texture and viscosity and increase the shelf life of meat products and prevents microbial spoilage. Sodium chloride reduction can be compensated by the addition of other salts with antimicrobial activity such as potassium chloride, magnesium chloride, potassium lactate or sodium diacetate [79, 86, 87]. Sodium chloride was successfully replaced in meat products without compromising the sensory characteristics by: potassium chloride (up to a ratio of 50:50 sodium chloride/potassium chloride) [85], as mixture of potassium chloride, magnesium chloride and calcium chloride (40–50% of sodium chloride replaced) [87], by calcium ascorbate or magnesium chloride improving the nutritive value of the product [88, 89]. To improve the taste of meat products, when sodium chloride content is reduced, the industry uses salt mixtures and bitter blockers such as adenosine monophosphate (AMP) or flavour enhancers like: monosodium glutamate, magnesium glutamate, yeast extracts or naringin [90, 91]. Phosphates can be used to reduce salt content as they promote protein extraction and increase water holding capacity in meat products [78].

Salt in Dairy Products

In cheese, salt creates flavours during ripening; determines moisture and texture of the cheese; regulates proteolysis; affects sugar fermentation, thereby controlling acidity; and ensures food safety and shelf life of the cheese [79, 92]. Salt reduction in cheese is possible by increasing the moisture content and decreasing the fat content [93]. Producing low-fat, low-sodium cheese is a challenge for the industry since it is strongly affects the taste [94]. The exclusive use of potassium chloride, magnesium chloride or calcium chloride (1.5%) as well as a 1:1 ratio of sodium chloride with magnesium chloride or calcium chloride resulted in unacceptable cheese [95, 96]. Only a mixture of sodium chloride and potassium chloride (1:1) achieved acceptable results in the tested products [95]. Another approach is the use of potassium-based emulsifying salts such as dipotassium phosphate and tripotassium citrate which can replace disodium phosphate and trisodium citrate and therefore reducing the sodium content of the cheese. However, this approach causes a loss of firmness of the processed cheese and has a strong impact on flavour due to potassium's metallic note [97]. Salt reduction in cheese presents a greater challenge than in bread and most meat products [75].

Sugar

Sugars are defined as either ‘extrinsic’ or ‘intrinsic’ depending on their origin. The term ‘sugars’ on back-of-pack labels is often used as a short form of ‘total sugars,’ whether extrinsic/added sugars or intrinsic/naturally present [98]. There are two group of sweeteners used to replace sugars in foods as a way to reduce the total sugar content by food reformulation. These are *bulk sweeteners* (used in large quantities) and *intense sweeteners* (used in very small quantities), they replace sugars in energy reduced foods or foods with no added sugars [98].

Bulk Sweeteners

Polyols are compounds derived from sugars. Polyols are bulk sweeteners which can be used to replace sugars in foods. Some polyols occur naturally e.g., sorbitol in fruits, but the industrial grades for production are made by hydrogenation of sugars. Therefore they are also often referred to as sugar alcohols, and are the following: *lactitol, isomalt, sorbitols, mannitols, maltitols, erythritol, and xylitol.* With the exception of erythritol, which is 0 kJ/g (0 kcal/g), the rest of polyols calorific value is approximately 10 kJ/g (2.4 kcal/g). Therefore polyols can be used to achieve a significant reduction in calorie content when used at high levels to replace sugars, being often used in combination with high-intensity sweeteners. The use of polyols in products is often kept to below 10% in the formulation, so that there will be no negative effects and impact on consumer acceptance [98]. However, their consumption in high quantities may cause gastrointestinal disturbances and it is frequently linked also to irritable bowel syndrome and abnormal flatulence [99, 100].

Intense Sweeteners

The permitted intense sweeteners are acesulfame K, aspartame, cyclamic acid and its sodium and calcium salts, saccharin and its sodium, potassium and calcium salts, sucralose, thaumatin, neohesperidine dihydrochalcone, steviol glycosides, neotame, advantame [98]. The use of these intense sweeteners is regulated in Europe, each with a maximum allowed dose. Therefore this needs to be considered when using them in products and in labelling of products [101, 102].

Other Sweeteners

Ingredients with bulking properties such as soluble dietary fibres and dextrins (inulin, oligofructose, polydextrose) are commonly used to replace sugars in formulations and to achieve a reduction in calories. These ingredients provide bulk and can have an additional positive effect of improving the nutritional content by increasing the dietary fibre content of products whilst reducing levels of sugars.

Sugar in Soft Drinks

The sugar functionality in soft drinks is diverse, such as providing sweetness, preservation, contribute to product colour, texture or mouthfeel. The ingredients (calorie-contributing) commonly used in the soft drink industry to provide sweetness are: sucrose, glucose/dextrose, high fructose corn syrup, invert sugar, glucose syrup/dried glucose syrup, fructose syrup [98]. Products that are 'sugar free' and 'reduced sugar' are the main options for reducing sugars in soft drinks. A number of sweeteners and blends are used to reduce the sugar content of soft drinks including acesulfame K, aspartame, aspartame-acesulfame salt, saccharin and its salts, saccharin sucralose, intense sweetener blends, sucrose/fructose/ high fructose corn syrup, glucose/intense sweetener, sucrose/intense sweetener [165].

Sugar in Dairy Products

The total sugars present in dairy products are usually a combination of intrinsic and added sugars (often added to increase sweetness). Lactose, a disaccharide formed of glucose and galactose units, is the only intrinsic sugar present in milk. Sugars have a functional role in dairy products including control of ice crystal formation, texture, water activity, sweetness, viscosity, stability to heat, supply bulk for low fat products fats, stabilizers for protein foams. Therefore, acesulfame K and sucralose, acesulfame K with aspartame and/or sucralose, small quantities of oligosaccharides, stevia and isomaltulose are among the sugar replacements used in reformulation of yogurts. The polydextrose or polyols, acesulfame K, sucralose and neotame, in combination with polyols, and fructose were successfully used to reduce the sugar content in ice cream and frozen desserts [98].

Sugars in Bakery Products

The bakery products category is quite varied covering many different types of products, with the level of sugar varying from 3% in bread up to 45% in iced biscuits. The main form of added (extrinsic) sugars in bakery products is sucrose, but also the invert sugar, brown sugar, molasses, fructose, starch hydrolysates, honey and malt syrup are used. In baking products, sugars influence many properties such as the volume, moistness, tenderness, colour, appearance, texture, viscosity, preservation and water activity, solubility, freeze/thaw stability, and energy content of baked products. Therefore, when reducing the sugars in bakery products, it is important not only to address the reduction in sweetness of the reformulated products. The following sugar substitutes were used for reducing and replacing the sugars in bakery products: polyols including lactitol, erythritol, maltitol and xylitol, polydextrose, gums, or modified starch (for texture, addition of bulking agents); crystalline maltitol (high sweetness and excellent humectancy); lactitol, polyol derived from lactose is stable at high temperatures and over a wide range of pHs; xylitol also presents good humectant properties, which is desirable for obtaining moist sponges and muffins [98].

Fat

World Health Organization recommends that the fat intake does not exceed 30% of total calorie intake; to limit intake of saturated fatty acids, which should not exceed 10% of total energy intake; and to consume less than 300 g of cholesterol daily [103]. High fat intake is associated with increased risk for obesity and some types of cancer, and saturated fat intake is associated with high blood cholesterol and coronary heart disease [104, 105]. Fat provides calories, fat-soluble vitamins, taste, texture and volume [106]. Fat can be replaced in foods by reformulating the foods with lipid-, protein-, or carbohydrate-based ingredients, individually or in combination.

Fat replacers represent a variety of chemical types with diverse functional and sensory properties and physiological effects which are categorised in *fat substitutes* and *fat mimetics* [107].

Fat Substitutes are mainly lipid-based macromolecules that are similar to fats and oils both physically and/or chemically such as: sucrose fatty acid esters and polyesters, carbohydrate fatty acid esters, various emulsifiers (such as mono- and diglycerides, lecithin), and structured lipids (such as medium-chain triacylglycerols) [108].

Sucrose Fatty Acid Polyesters are mixtures of sucrose esters formed by chemical transesterification or interesterification of sucrose with six to eight fatty acids. They are manufactured from saturated and unsaturated fatty acids of chain length C12 and higher, obtained from conventional edible fats and vegetable oils [109-112].

Sucrose Fatty Acid Esters are mono-, di-, and tri-esters of sucrose with fatty acids, made in a manner similar to sucrose polyester [113]. They are used as emulsifiers, stabilisers, lubricants, anticaking agents, thinning agents, and antimicrobials [114-117].

Polyol Fatty Acid Esters are prepared by reacting one or more fatty acid esters with a polyol containing at least four hydroxy groups in the presence of a basic catalyst [117]. Examples include sorbitol, trehalose, raffinose, and stachyose polyesters [109].

Structured Lipids are triglycerides containing short chain fatty acids and/or medium chain fatty acids and long chain fatty acids. They are prepared by chemical and enzymatic synthesis or random transesterification [119-121]. The medium chain triglycerides contain saturated fatty acids of chain length C8:0 (caprylic) to C10:0 (capric) with traces of C6:0 and C12:0 fatty acids. They are manufactured from vegetable oils, such as coconut and palm kernel oils, by hydrolysis followed by fractionation of the resulting fatty acids to concentrate C8 and C10 fatty acids, and re-esterification with glycerol to form triglycerides [122-124]. The short and long acyl triglyceride molecule are structured triglycerides comprised of a mixture containing at least one short chain fatty acid (primarily C2:0, C3:0, or C4:0 fatty acids) and at least one long chain fatty acid (predominantly C18:0, stearic acid) randomly attached to the glycerol backbone. They are used as chocolate-flavoured coatings, deposited chips, caramels and toffees, fillings and inclusions for confectionery and baked goods, peanut spreads, savoury dressings, dips and sauces, and dairy products such as sour cream, frozen dairy desserts, and cheese [125].

Esterified Propoxylated Glycerols comprise a family of derivatives of propylene oxide, synthesized by reacting glycerol with propylene oxide to form a polyether polyol that is subsequently esterified with fatty acids. They are used as replacement for fats and oils in a variety of products including frozen desserts, salad dressings, baked goods, and spreads and for cooking and frying.

Trialkoxytricarballylate, *Trialkoxycitrate*, and *Trialkoxyglyceryl Ether* are polycarboxylic acids with two to four carboxylic acid groups esterified with saturated or unsaturated alcohols having straight or branched chains of 8–30 carbon atoms [126].

Fat mimetics are normally carbohydrate-based compounds (modified starches and hydrocolloids) or protein macromolecules that are mimic the organoleptic and physical properties of fats mainly by water binding. Fat mimetics are generally carbohydrates or proteins exhibiting polar and water-soluble properties, which create a creaminess and lubricity state similar to full-fat products. Fat mimetics are not able to replace the non-polar functional properties of fat, like the flavor carrying ability. The main food industry applications where fat replacers are used to reduce/reformulate the fat content of the food are the following: bakery, frying, frozen desserts, margarine, butter and spreads, confectionary, processed meat, dairy and other processed foods [107]. In these applications the fat replacer could be lipid based and acting as emulsifier, providing cohesiveness, tenderizer, carry flavour, replace shortening, prevent staling, prevent starch retrogradation, condition dough, texturiser, spreadability, flavour, plasticity; carbohydrate based, used for moisture retention, staling retardation, texturiser, thickener, increasing viscosity, increase the water holding capacity, texturize, provide mouthfeel; or could be protein based, acting as mainly as texturiser, also as stabiliser, emulsifier, water holding, and providing mouthfeel [107].

Protein-Based Fat Mimetics are fat replacers that are derived from a variety of protein sources, including egg, milk, whey, soy, gelatine, and wheat gluten. Some of these protein-based fat mimetics are microparticulated to form microscopic coagulated round deformable particles that mimic the mouthfeel and texture of fat. They are generally used in dairy products, salad dressings, frozen desserts, and margarines.

Carbohydrate-Based Fat Mimetics serve as thickeners or gelling agents in foods. Gums, starches, pectin, cellulose, and other carbohydrate ingredients provide some of the functions of fat in foods by binding water. They also provide texture, mouthfeel, and opacity [127]. Corn syrups, syrup solids, and high-fructose corn syrups are used as fat replacers cookies to control water activity. Microcrystalline cellulose mimics fat in aqueous systems, contributing to consistency, and mouthfeel; stabilizes emulsions and foams; and add viscosity, gloss, and opacity to foods.

Powdered cellulose is also effective in reducing the fat in fried batter coatings and fried cakes [118].

Maltodextrins are mixtures of saccharide polymers of varying chain lengths, being are produced by partial hydrolysis of starch obtained from corn or potato starch. Maltodextrins are used to build solids and viscosity, bind/control water, and contribute smooth mouthfeel in fat replacing systems for table spreads, margarine, imitation sour cream, salad dressings, baked goods, frostings, fillings, sauces, processed meat and frozen desserts [107].

Polydextrose is a randomly-bonded polymer of glucose, sorbitol, and citric or phosphoric acid. It is used as fat mimetics in many food categories, including baked goods and baking mixes, chewing gum, confections and frostings, salad dressings, frozen dairy desserts and mixes, gelatins, puddings and fillings, hard and soft candy, peanut spreads, fruit spreads, sweet sauces, toppings, and syrups [107].

Fortification of Foods

According to WHO and FAO the fortification is defined as "the practice of deliberately increasing the content of an essential micronutrient, i.e., vitamins and minerals (including trace elements) in a food, so as to improve the nutritional quality of the food supply and provide a public health benefit with minimal risk to health" [129]. Worldwide, the three most common forms of micronutrient malnutrition are *iron, vitamin A and iodine deficiency*. In total, these affect at least one third of the world's population, the majority of whom are in developing countries [129]. The food-based strategies to increase the micronutrients intake are identified as the following: dietary diversification and food fortification, as well as nutrition education, public health and food safety measures, and finally supplementation. Of these, food fortification tends to have less immediate but nevertheless a much wider and more sustained impact [129].

Iron Fortification

A wide variety of iron compounds are currently used as food fortificants and these are listed in Table 4 [130-132].

Table 4. Chemical compounds used for food fortification with iron

Water Soluble Fortificants	**Poorly Water Soluble Fortificants***	**Water Insoluble Fortificants**
ferrous sulfate heptahydrate ferrous gluconate ferrous lactate ferrous biglycinate ferrous ammonium citrate sodium iron EDTA	ferrous fumarate ferrous succinate ferrous saccharate	ferric orthophosphate ferric pyrophosphate elemental iron

*soluble in dilute acid.

For most food vehicles, the recommended iron fortificants, in order of preference, are: ferrous sulphate > ferrous fumarate > encapsulated ferrous sulphate or fumarate > electrolytic iron (at twice the amount) > ferric pyrophosphate (at twice the amount). The addition of ascorbic acid (vitamin C) or sodium ethylenediaminetetraacetic acid (sodium EDTA) and the removal of phytates can be effective ways of increasing the total amount of iron absorbed from iron-fortified foods. The co-addition of ascorbic acid in a 2:1 molar ratio is recommended in order to enhance iron absorption. This applies mainly to infant foods. In the case of high phytic acid containing foods, the molar ratio (ascorbic acid: iron) can be increased to 4:1. Novel iron fortificants developed are: sodium iron EDTA, ferrous bisglycinate, encapsulated forms, with ferrous sulphate and ferrous fumarate, and micronized ferric pyrophosphate [132].

Iron Fortification of Specific Foods

In Latin America, more than 20 countries, have implemented mass iron fortification programmes, most of which involve the fortification of wheat or maize flours (). Elsewhere, other frequently used food vehicles include cereal-based complementary foods, fish sauce, soy sauce and milk. Salt has also been fortified with iron in efficacy trials. Products derived from cereal flours (e.g., bread, cereal snacks and breakfast cereals) are also useful food vehicles, but the amount of iron provided via this route will depend on the quantity of food eaten and on the level of fortification. The use of sodium iron EDTA is recommended for the mass fortification of high-phytate cereal flours and for sauces with a high peptide content (e.g., fish sauce, soy sauce). For liquid milk products, ferrous bisglycinate, micronized ferric pyrophosphate and ferric ammonium citrate are the most appropriate fortificants [129].

Vitamin A and β-Carotene Fortification

Vitamin A is fat-soluble, therefore can be easily added to fat-based or oily foods. According to WHO and FAO, Vitamin A fortificants can be divided into two categories: (1) oily forms that can be incorporated directly into fat-based foods or emulsified into water-based ones (e.g., milk); and (2) dry forms that can be dry mixed into foods or dispersed in water, depending on whether they are cold water dispersible or non-cold water dispersible [132]. The commercially available forms of vitamin A are the following: oily vitamin A acetate (retinol ester of acetic acid stabilised with antioxidants), oily vitamin A palmitate (retinol ester of palmitic acid stabilised with antioxidants), oily vitamin A palmitate or acetate with vitamin D_3 (retinol ester and colecalciferol mix stabilised with antioxidants), dry vitamin A palmitate or acetate (vitamin A embedded in water soluble matrix and stabilised with antioxidants), dry vitamin A palmitate or acetate with vitamin D_3 (vitamin A and vitamin D_3 embedded in water-soluble matrix and stabilised with antioxidants [133, 136]. Through various fortification programs in South America and Asia, the main foods used for vitamin A fortification were: margarines, vegetables oils, hydrogenated fats, wheat flour, maize flour and sugar [134]. A variety of oily and dry forms of the retinol esters, retinyl acetate and retinyl palmitate, are available for food fortification purposes. The dry forms are usually gelatin-, starch- or gum-coated and all forms contain antioxidants. Vitamin A fortification of margarine and sugar has been shown to be efficacious. Vegetable oils and cereal flours are also considered to be useful fortification vehicles [129]. Other foods that have been fortified successfully with preformed or provitamin A include: dry milk, complementary foods for infants and young children, biscuits and beverages, which are sold commercially or used in school feeding programmes (in Indonesia, Mexico and Central America [135,136], Peru [137] and South Africa [138], instant noodles (in Thailand), [139] and yoghurt (worldwide) [140].

Iodine Fortification

WHO and FAO recognise two chemical forms of iodine for use as food fortificants, namely, iodate and iodide, both usually added as the potassium salt, but sometimes as the calcium or sodium salt [129]. Potassium iodate is preferred to potassium iodide for salt iodization because it is more stable. The iodization of all salt for both human and animal consumption, is the strategy recommended by WHO to correct iodine deficiency. Potassium iodide has been used as an additive in bread and salt for about 80 years, and potassium iodate for about 50 years. Iodates are less soluble in water than the iodides, more resistant to oxidation and evaporation [129]. Other food vehicles used for iodine fortification are: bread (e.g., Russia, where it is a staple food, [141, 142], milk (in Europe and UK) [143] and water (e.g., Thailand) [144].

Food Fortification with Bioactives by Encapsulation

Encapsulation may be defined as a process to entrap one substance within another substance, thereby producing particles with diameters of a few nanometers to a few millimeters [145]. Microencapsulation has been widely used for many industrial applications including food, pharmaceutical, and agricultural. It is a process of enclosing micron-size core particles of solids, and droplets of liquids or gases in an inert shell. The inert shell in turn isolates and protects the enclosed core materials from the external environment [146]. The application of microencapsulation has been extended to the food industry, mainly for controlling the release of flavorings and the production of foods containing functional ingredients (e.g., probiotics and bioactive ingredients). Microencapsulation techniques are divided into three classes: chemical, physical, and physico-chemical methods [147]. Chemical methods include *in situ* polymerization and use of liposomes; physical methods include spray-drying and fluidized bed coating, and physico-chemical methods include coacervation and sol-gel encapsulation [148, 149]. The microcapsules produced by each method are different [150]. According to Đorđević, et al., the encapsulation processes for food applications are classified in *spraying technologies*, and *microgel encapsulation*. Spraying technologies are: *spray drying*, where commonly encapsulated bioactives are hydrophobic compounds and lyophilised proteins; *spray cooling/chilling*, mainly for hydrophobic compounds, heat sensitive bioactives; *fluidized bed coating*, used for encapsulation of solid-core materials. Microgel encapsulation technologies are: *Extrusion process*, are mainly used for encapsulation of cells, plant bioactives, enzymes, and proteins; *Emulsification process*, used for encapsulation of hydrophilic and hydrophobic/lipophilic compounds; *Complex coacervation,* used for hydrophobic compounds, flavours, essential oils; *Emulsion technologies,* for hydrophilic and hydrophobic/lipophilic compounds; *Encapsulation in liposomes,* for hydrophilic and hydrophobic/lipophilic compounds, Enzymes; *Encapsulation in cyclodextrins,* used for apolar molecules, and flavours; *Supercritical fluid-based encapsulation,* for hydrophilic and hydrophobic/lipophilic compounds; and *Encapsulation in yeast,* used for aromas and antioxidants [145].

The key criteria in selecting a suitable microencapsulation technology is to select a process that can meet the product specifications at the minimum cost, reproducibility, available facilities, and high throughput [147]. Coating materials used in microencapsulation mainly fall into the following categories: water-soluble polymers, water-insoluble polymers, water-soluble non-polymeric material and water-insoluble non-polymeric material. The water soluble materials for coating are cellulosic polysaccharides (i.e., methylcellulose, hydroxypropyl cellulose), non-cellulosic polysaccharides (i.e., guar gum, chitosan, sodium alginate) and non-polysaccharides (i.e., gelatine, polyethylene glycol, polyvinyl alcohol). The non-water soluble coating materials are: the organic soluble polymers (i.e., ethylcellulose, shellac, methylacrylates, cellulose acetate butyrate, cellulose acetate phthalate); latexes, dispersions, pseudolatexes (i.e., methacrylic acid co-ethyl acrylate, ethylcellulose, cellulose acetate phthalate); and non-polymers (i.e., waxes, sugar, sorbitan esters, and linear fatty acids) [147].

Applications of Encapsulation for Food Reformulation

Spray drying techniques are most commonly used in the food industry. They were successfully applied for the encapsulation of flavours (i.e., oregano, citronella), ethyl butyrate and caprylate, milk fat mainly using whey protein [151-153]. These techniques were also used for the encapsulation of soya oil, calcium citrate and lactate, lycopene, fish oil, arachidonyl l-ascorbate, cardamom oleoresin and essential oil, d-limonene, l-menthol, short chain fatty acid [154-163]. In the food industry the preparation of ingredients that have functional health benefits, such as antioxidants and probiotics, are of great interest [164].

Beverages

Burin et al., 2011, [165], evaluated the stability of anthocyanin, which was encapsulated within different carrier agents in an isotonic soft drink system. Microencapsulation has been used to increase the stability of these compounds [166]. The spray-drying technique was used to encapsulate anthocyanins originated from Cabernet Sauvignon grapes. Aditya et al., 2015, [167] prepared microcapsules of curcumin and catechin using water-in-oil-in-water emulsions since they are unstable compounds and easily degraded in the presence of oxygen, alkaline pH, and high temperature [168]. Kausadikar et al., 2015, [169] encapsulated lemon oil with maltodextrin by the spray-drying technique to prevent oxidation during storage, therefore the encapsulated lemon oil can be stored for six months.

Baked Goods

Rocha et al., 2012, [170] produced microcapsules of lycopene by spray-drying, using a modified starch as the encapsulating agent since this is easily decomposed by oxidation during the storage process, due to its high number of conjugated double bonds. O'Brien et al., 2003, [171] encapsulated vegetable shortening to increase oxidative stability and convert fat into a stable powder for use in short dough biscuit production. The microencapsulated high-fat powders could be used as a replacement for fat/oil in commercial biscuit production [167].

Meat and Poultry

Since probiotic organisms have poor survival in fermented foods [172-174], to improve the viability of bacterial cells, microencapsulation technique was used to retain them within a protective polymer membrane or matrix *Lactobacillus reuteri* could be used in dry fermented foods, after the microencapsulation, which can prevent the loss of cell viability after drying without altering the sensory quality of products [175]. Muthukumarasamy and Holley, 2006, [176] used probiotics in dry fermented sausages, to increase the nutritional value. In order to improve the oxidative stability and sensory properties of chicken nuggets during frozen storage, Jiménez-Martín et al., 2016, [177] used microcapsules of omega-3 fatty acids from fish oil for enrichment. Comunian et al., 2014, [178] encapsulated ascorbic acid to improve on physicochemical and sensory stability of chicken frankfurters.

Dairy Products

Anjani et al., 2007, [179] developed microcapsules of Flavourzyme (a commercial peptidase preparation from Aspergillus) with various materials, for use in cheese production. These microcapsules added to milk during cheese manufacturing allowed for good distribution of Flavourzyme. Champagne et al., 2015, [180] evaluated the effects of microencapsulation by the spray-drying technique on the stability of probiotic bacteria in ice cream, the microencapsulated bacteria having higher survival rates compared to the non-encapsulated culture.

The food industry has a big role in helping consumers eat nutritious, healthily and sustainably, and the food reformulation is a major feature of this role. With reformulation, the industry need to address huge current issues related to *health and nutrition of the population* (i.e., obesity, malnutrition), *environmental impact* (i.e., climate change, waste, sustainable raw materials), *economic and energy-efficiency* (i.e., choice of equipment and technologies). However, in order to address all these challenges, reformulation cannot act alone, as unique solution; with many technical difficulties faced by the food reformulation and with its complex approaches it is only part of the solution.

Improving Economic by Exploiting the Potential of Co-Products

The food industry produces high quality, safe and nutritious foods. However, in addition there are a number of components that are not part of the primary production stream but which can be processed to provide added value products.

Recovering Dietary Fibre from Food Co-Products

One major driver for the development of new foods from by-products has been in the development of those containing functional compounds with impact on human health. In particular, a group of compounds that has become important because of potential health benefits, but also in the applications on the food industry, are those that constitute dietary fibre

[181]. A number of health claims [7] have been approved by EFSA for fibre including increasing faecal bulk and reduction of post-prandial glycemic response. However the majority of fibre related claims [40] have been rejected. This latter group includes claims related to weight management, control of cholesterol levels, healthy digestion and the effect on the gut microflora. Other components such as oat beta glucan and resistant starch do have approved health claims.

The CODEX *Alimentarius Commission* (CAC) in 2009 defined dietary fibre as carbohydrate polymers with ten or more monomeric units (that may include compounds with degree of polymerization 3–9), which are not hydrolyzed by the endogenous enzymes in the small intestine of humans and belong to: edible carbohydrate polymers naturally occurring in the food as consumed, carbohydrate polymers obtained from raw materials by physical, enzymatic or chemical means and synthetic carbohydrate polymers with proven health promoting physiological benefits [182]. This definition also includes resistant oligosaccharides, resistant starches and resistant maltodextrins [182].

Chemical, mechanical, thermal and/or enzymatic treatments have been used to extract dietary fibre from different vegetable sources, with the aim of modulating the functional properties of the polysaccharides obtained. Therefore, in order to obtain products with the desired yield and functional properties, research is focussed to optimise conditions such as temperature, extraction time, pH, solvents [181]. Due to the complex composition of the polysaccharides, the extraction process can be challenging and mechanical and enzymatic pre-treatments are often used to improve extractability. Hot water extraction followed by chemical treatment are the most common methods used, but, new alternative extraction methods use enzyme, combined methods or assisted treatments [183-187].

Thermal Extractions

Generally, thermal processes break the relatively weak glycosidic bonds within the polysaccharides structure causing the cell to swell and break, which modifies the polysaccharide properties [188]. Thermal extraction is one of the most common treatments and has been used for several years with high extraction yields, principally of soluble dietary fibre such as pectin. Generally, during thermal extractions the temperature increase can lead to loss of product quality and so complementary treatments such as acidified or alkaline media, solvents or chelating agents and even the use of microwave are employed to help the release of the polysaccharides from the cell wall [181]. Thermal extractions are widely used for extraction of dietary fibre from fruit by-products, like the extraction of pectins from orange peel (189), from dried orange albedo and peel [190], apple pomace and citrus peel [191].

Chemical Extractions

Chemical extraction has been frequently used either alone or in combination with high temperatures and long processing times under acid or alkali treatments. Common solvents used for polysaccharide extraction are dimethyl sulfoxide, urea, hydrochloric acid, potassium, sodium and barium hydroxide [192]. Alkaline treatment being the most frequently used [193] and extraction conditions differ depending on the food matrix and the polysaccharide that is going to be extracted. Alkaline extraction releases various polysaccharides from the cell wall due to the disruption of both hydrogen and covalent bonds. The hydroxyl ions disrupt the hydrogen bonds and break ester linkages between cellulose and hemicellulose solubilizing part of the material. Alkaline extractions have been used for polysaccharides such as heteroxylans associated from maize bran [195] and fibre from linseed meal [196] and pea hulls [197]. Under acid extraction conditions, hydrolysis may occur [194]. Acidic condition have been used for fibre extraction from fractions obtained after protein removal from soybean [198].

In general, both alkaline and acid treatment methods use high temperatures, whereas the usage of solvents, oxidants, chelating agents is performed at ambient temperatures. However all of them require long periods of time [189, 199-201].

Enzymatic Extractions

Enzymatic treatments are used to overcome some of the disadvantages of chemical or thermal methods, such as high temperatures, low pH, corrosion and large volumes of waste that needs neutralization [197]. The use of enzyme treatment results in applying extraction conditions similar to those encountered in the human digestive system. Thus many of the enzymes extractions of dietary fibre used biological molecules for *in vitro* digestive protocols are digestive enzymes such as pancreatin, pepsin, trypsin, chymotrypsin, α-amylase, lipase [181]. Enzymatic extraction was successfully used for a range of fibre extraction protocols from fruit, vegetable and cereal co-products, such as oat bran (using, amylases, proteases and lipase) [202] okra by-product [203], starch free potato pulp [204], rapeseed cake [205], lime peel [206], apple pomace [197], and pod waste [207].

Emerging Technologies Assisted Extraction

The thermal, chemical and enzymatic extraction processes mentioned can be combined or used with other technologies such microwave, ultrasound and high-pressure to optimize dietary fibre extraction. In order to develop new processes with promising results on dietary fibre functionality, efforts and more research needs to be done on synergies between these technologies and the more established chemical, thermal and enzymatic treatments [181].

Recovering Protein from Food Co-Products

Protein Recovery from Co-Products of Plant Origin

Many methods and processes of protein purification by removal of non-protein components or protein recovery by extraction and precipitation have been developed for the production of protein-rich products [208]. Following methods developed for soya beans, similar technologies then were applied to various by-products resulting from oil extraction (oilseeds or meals), to legumes and cereals and products obtained from tomato or fruit juices processing [209, 210]. The protein products are categorized based on their protein content as protein flours, concentrates, and isolates with about 50%, 70%, and higher than 95% protein content, respectively. Main raw materials for protein products are oilseeds and legumes [211] with soya protein products being the main type currently produced [212].

Protein Flour

Protein flours and grits are products directly resulting from industrial oil processes. The oil is conventionally solvent extracted by hexane, or mechanical pressed before solvent extraction. In the case of food grade flour production, before oil extraction, the oilseeds must be purified from foreign materials, impurities, and hulls by passing through scalpers and aspirators, then dried and tempered. The seeds are then cracked in roll mills and flaked for the oil to be extracted. After solvent extraction the defatted flakes are desolventized [209, 211, 213, 214]. The typical composition of defatted soy flour is 53% protein and less than 1% fat, while that of unprocessed flour is 41.5% protein and 18–21% fat [212].

Protein Concentrate

Protein concentrate (70% protein) is obtained after removing carbohydrates, salts, and other low-molecular-weight components from the defatted flakes. The main extraction method for preparation of protein concentrates are the use of water–alcohol (methanol, ethanol, or isopropanol aqueous solutions), [211, 215, 216]; of dilute acid solution (at low pH, i.e., for soya at pH 4.5), [209, 211, 217]; and hot water after protein denaturation (66–93°C, at appropriately adjusted pH, i.e., for soya bean in the range of pH 5.0–7.5), [211, 217]. Membrane technology has also been developed for protein concentrate production, using ultrafiltration and membranes with 100,000 molecular weight cut off (MWCO). In this way protein is separated from low molecular-weight substances (salts, carbohydrates, peptides, colorants, odorous materials) and the protein concentrate is obtained by spray-drying [218, 219].

Protein Isolates

The protein isolation technology was first applied to soya proteins. Protein content in the protein isolates is more than 90%, and they have characteristic properties. The protein is typically extracted, precipitated and dried. The extraction and precipitation are significant for the efficient isolation of proteins since they are directly linked with the protein solubility [220-222]. The temperature, pH, agitation, liquid-to-solids ratio, particle size of ground solid material, flakes formation, and extraction time are main factors that influence the protein extraction. The extraction is usually performed with alkaline water at pH 8–10 for most of oilseeds proteins, at elevated temperatures of 50–55°C. Various salt solutions also have been used in the protein extraction of oilseed meals such as sodium sulfite and chloride with sunflower seed and cottonseed [223, 224], sodium carbonate and bicarbonate with sunflower seed [225], calcium carbonate, magnesium and sodium sulphate in cottonseed [226], hexametaphosphate with flaxseed [227], and sodium chloride and carbonate with the tomato seed or meal [228-230]. The most common liquid-to-solids ratio of 1:10–1:20 and the agitation and extraction time of 45–60 minutes are adequate to maximize the yield of protein extraction [212]. After extraction, the extract is separated from the solids by screening, centrifugation or decanting, and filtration [211, 231]. The proteins contained in the extract will be recovered by precipitation. For the precipitation of the protein from the extract, can be used various approaches such as including organic solvents, divalent cations, heat, acid-bases (for pH adjustment), salts, nonionic polymers, and polyelectrolytes [232-235] A common method which is used for precipitation uses acids (i.e., acetic, hydrochloric, phosphoric, or sulfuric acid) at the isoelectric point of proteins [237, 238]. The protein curd is collected by centrifugation or filtration, washed, and dried (i.e., spray-dried). Membranes of 10,000–100,000 MWCO are used for the concentration of the protein extract followed by drying [236].

Protein Recovery from Co-Products of Animal Origin

Fish by-products are good sources of minerals, proteins and fat. Fish by-products from various marine species have been assessed for their potential use in production of fish protein hydrolysate by enzyme treatment, and use as a food ingredient, [239, 240]. Recovery of proteins from fish waste products by alkaline extraction resulted in good yield of protein [241]. The isoelectric solubilisation/precipitation of fish muscle proteins with concurrent separation of lipids was also efficiently used for protein recovery from fish by-products [242-245]. In fish muscle homogenates, proteins are present as aggregates that are held by weak protein–protein hydrophobic interactions [246]. The solubility of fish muscle proteins can be influenced by providing conditions which either favour or disfavour protein solubility. When acid is added protonation of negatively charged side chains on glutamyl or aspartyl residues results in an increased net positive surface charge. When base is added, deprotonation of the side chains on tyrosyl, tryptophanyl, cysteinyl, lysyl, argininyl or histidinyl residues contributes to an increased net negative surface charge [247]. Therefore, the solubilisation of fish muscle protein is mainly attributed to the protonation of aspartyl and glutamyl residues at acid pH and deprotonation of lysyl, tyrosyl, and cysteinyl at alkaline pH. It is possible to adjust the pH of a protein solution so that the number of negative charges on a protein's surface is equal to the number of positive charges; and therefore, the protein molecule assumes a zero net electrostatic charge. The pH at which the net electrostatic charge of a protein is equal to zero is called the isoelectric point [247]. The hydrophobicity-driven protein–protein interactions are favored at the isoelectric point; and therefore, proteins at their isoelectric point achieve minimum solubility and typically precipitate.

The second major group of fish muscle proteins, sarcoplasmic proteins are water soluble at both acidic and basic pH [247]. The isoelectric behaviour of fish muscle proteins can be used to recover functional proteins from aquatic animals processing by-products as well as low-value aquatic species. Since muscle proteins are in a soluble form, the insoluble components (bone, skin, scale, etc.) can therefore be removed from the solution by, for example centrifugation, followed by protein precipitation and recovery at the isoelectric point. This technique has also been successfully applied to processing various fish by-products [248, 249], krill [250], and whole fish [249, 250]. Additionally, the technique has been used to recover proteins and lipids from beef processing by-products [251], mechanically deboned turkey [252] and chicken [253].

Phytochemical Recovery from Food By-Products

Fruit, vegetable, and oilseed processing result in various amounts of by-products. The plant based by-products materials mostly used are generated from citrus, carrot, grape, peach and apricot, apple, grapes, passion fruit, sugar beet, sugarcane, banana, kiwifruit, potato, and cereals [254-258]. Main phytochemicals which are commonly recovered from plant based by-products are: carotenoids and phenolic compounds (i.e., tocopherols; anthocyanins, flavonoids and related compounds like coumarins, cinnamic acid derivatives, and chalcones; phenolic diterpenes; and phenolic acids).

The solid waste, often called "pomace," is obtained by pressing fruits or vegetables and can contain pulp, peels, seeds and, stones [259] It is considered a rich source of phenolics including flavanols (catechin, epicatechin, procyanidins), flavonols, hydroxycinnamates, and dihydrochalcones [260]. Organic solvents are commonly used for extraction [261, 262]; acetone being the most efficient when compared to other solvents like ethanol, petroleum ether, and hexane [263]. When fresh, wet raw material is used for carotenoids recovery, an initial step of water washing is necessary to remove most free sugars and other soluble compounds. The initial washing step is followed by washing with acetone, to remove the water and facilitate the carotenoid extraction in the following steps. Successive extractions are needed for good carotenoid quantitative recovery. Therefore, two extractions with acetone, (solvent-to-solid ratio of 2:1) at ambient temperature, recovers around 90% of the total carotenoids from orange peel [263]. Maceration of the raw material by enzymic treatment increases the yield [264], but may affect the quality of other biomolecules, which can be recovered such as pectin. The extracted carotenoids can be obtained as a crude pigment, after solvent evaporation at low temperature, or a purification step can be carried out by solvent–solvent transfer to hexane [263]. Pigments from carrot pulp waste have been produced by spray drying or freeze drying. Freeze-dried powders had a better pigment stability than spray-dried ones, the stability also improves if appropriate packaging methods and storage conditions are used [265, 266]. Other colorants that can be recovered from plant by-products are anthocyanins from grape pomace [267] or banana bracts [268].

Anthocyanins, can be efficiently extracted from grape skins by heat treatment at 70 °C or by novel technologies such as ultrasonication, high hydrostatic pressure, and pulsed electric fields. The engagement of these technologies increased total phenolic content by 50% [269]. Co-products like grape skins and seeds [269-273], citrus peel and pulp [274-276], apple pomace [260, 277], potato peel waste [278, 279], onion by-products [280, 281], bean hulls [282, 283], etc., are rich in antioxidants. Oilseed meals remaining after oil extraction, wastes of olive oil production, and by-products from cereal processing are other potential sources of antioxidants [284, 285]. The antioxidant (total phenol) contents of the main by-products could vary, with red onion scale content of 105.5 g/kg dry matter (expressed as ferulic acid) to 0.6 g/kg in oat hulls [286, 287]. The amount of phenolics in plants also depends on many factors such as variety, harvest season, soil, geographical region and ultimately of the method of isolation and analysis used to measure them. Phenolic antioxidants are mainly isolated by solvent extraction. The quality and quantity of the molecules extracted is strongly dependent on the solvent, due to the variant antioxidant potential of compounds with different polarity. Nonpolar solvents (hexane, petroleum ether) can be used for the recovery of tocopherols and certain phenolic terpenes. Ethyl ether and ethyl acetate are very efficient for the recovery of flavonoid aglycones, low-molecular-weight phenols, and phenolic acids. Solvents of higher polarity (ethanol or ethanol–water mixtures) additionally can extract flavonoid glycosides and higher molecular weight phenols, resulting in higher yields of total extracted polyphenols [212]. Since, many undesirable substances are usually co-extracted, purification is necessary to isolate the antioxidant fraction. Other parameters that affects yield and antioxidant activity of the extract are temperature, time of the extraction and pH [284, 285]. Acidic conditions favour the extraction with aqueous or alcoholic medium for some raw materials, like grape by-products (pH 3–3.5) [288], olive rape (pH 4), or wheat bran [284], while alkaline pH results in maximum antioxidant activity of extracts from oat fibre [289]. Lignans were purified from flax seeds using aqueous methanol or ethanol after defatting with hexane, followed by alkaline treatment with sodium hydroxide [290].

A method was patented for the production of nutraceuticals from palm fatty acid distillate, involving treatments with alkyl alcohol and sodium methoxide for the conversion of free fatty acids and glycerides into alkyl esters. With the removal of the alkyl esters by distillation under reduced pressure, tocopherols, tocotrienols, sterols, and squalene remained in the residue. Following a cooling step to crystallize and separate the higher-melting sterols from tocopherols and tocotrienols, passage of the filtrate containing tocopherols, tocotrienols and squalene through an ion-exchange column with anionic exchange resin to remove the squalene, produced a concentrated tocopherol and tocotrienol fraction in the solvent which was then removed [291, 292]. Supercritical CO_2 has shown to be efficient for the purification of plant extracts [273, 285].

The agro-food industry faces critical global issues due to its exponential growth of by-products. The concerns are related to their storage, disposal, environmental impact, and potential health risks. However, the exploitation of by-products for the recovery of added-value compounds like macronutrients and phytochemicals could offer a sustainable avenue for industrial growth and waste management due to the fact that dietary bioactives (i.e., phytochemicals) are widely investigated in the field of chemistry, biology, nutrition, and medicine for their potential health-promoting effects. This can translate into the research and development of new functional foods and health products from low cost raw materials and it crucial for nutraceutical, cosmetic, pharmaceutical, and agribusiness sectors. These issues call for an imperative technical innovation or existing technological adjustment and new applications in order to optimize the processing methods of waste products to reduce biomass utilization and environmental risks as well as to improve recovery of added-value compounds. Ultimately, the utilization of food by-products for recovering nutraceuticals represents an economical and sustainable way of production and "waste" disposal and will be among the key strategies to combat food and nutrition security demands of the future.

CONCLUSION

It is essential to have a holistic and balanced view on food processing, taking on board positive and negative aspects. The management of risks to food safety and stability constitutes an essential element of food security which can be achieved through food processing. There needs to be a re-evaluation about how food processing can be better applied to create food products more efficiently, implying a better integration along the whole food supply chain, more energy and nutrient efficient, from the farm to the consumer, with attention to quality, sustainability, logistics, food products and processes is required.

Hajkowicz on 2015, [293], has identified seven global megatrends, which will influence how we deal with food and nutrition security across the food supply chain. These megatrends are: More from less, Planetary pushback, The silk highway, Forever young, Digital immersion, Porous boundaries and Great expectations. These megatrends are due to shifts in geopolitical, environmental, economic, social or technology conditions that substantially change the way people live, will shape our world in the next 20 years [293]. On 2016, Augustin et al., [294], translated Hajkowicz's global megatrends into food megatrends and opportunities for a future food supply chain as follows: More food from less resources, Foods for a healthy planet, Foods for the Asian century, Foods for beauty and health, Digital food and the internet of food, Producing food in a globally networked environment, Foods that meet our expectations [294]. It is very clearly identifiable that food manufacture, processing and reformulation is pivotal to deliver these food megatrends.

Engagement and effective communication between all stakeholders along the food supply chain, including consumers and government, is essential for delivering innovative solutions for food and nutrition security. There needs to be a closer integration between social science and the sciences that underpin innovations in technology to understand and respond to consumer concerns, opposition to technological solutions and address issues that arise in the complex food supply chain [294-296].

REFERENCES

[1] FAO 2008, *Food Security Information for Action Practical Guides*, An Introduction to the Basic Concepts of Food Security, www.foodsec.org/docs/concepts_guide.pdf.

[2] Froder, H., et al., Minimally processed vegetable salads: Microbial quality evaluation. *Journal of Food Protection*, 2007. 70(5): p. 1277–1280.

[3] Tournas, V. H., Moulds and yeasts in fresh and minimally processed vegetables, and sprouts. *International Journal of Food Microbiology*, 2005. 99(1): p. 71–77.

[4] Cliffe-Byrnes, V., and O'Beirne, D., Effects of chlorine treatment and packaging on the quality and shelf-life of modified atmosphere (MA) packaged coleslaw mix. *7th Karlsruhe Nutrition Congress on Food Safety. 2002*. (p. 707–716). Germany Elsevier Sci Ltd.

[5] Zagory, D., Effects of post-processing handling and packaging on microbial populations. *Postharvest Biology and Technology*, 1999. 15(3): p. 313–321.

[6] Arvanitoyannis, I., and Stratakos, A. C., Chapter 12—Irradiation of fruits and vegetables. *Irradiation of food commodities*. 2010. (p. 467–535). Boston Academic Press.

[7] Nguyenthe, C., and Carlin, F., The microbiology of minimally processed fresh fruits and vegetables. *Critical Reviews in Food Science and Nutrition,* 1994. 34(4): p. 371–401.

[8] Ramos, B., et al., Fresh fruits and vegetables—An overview on applied methodologies to improve its quality and safety, *Innovative Food Science and Emerging Technologies, Innovative Food Science and Emerging Technologies*, 2013. 20: p. 1–15.

[9] Artés, F., et al., Sustainable sanitation techniques for keeping quality and safety of fresh-cut plant commodities. *Postharvest Biology and Technology,* 2009. 51(3): p. 287–296.

[10] Bharathi, S., et al., Predicting the behavioural pattern of Escherichia coli in minimally processed vegetables. *Food Control*, 2001. 12(5): p. 275–284.

[11] Selma, M. V., et al., Effect of gaseous ozone and hot water on microbial and sensory quality of cantaloupe and potential transference of Escherichia coli O157:H7 during cutting. *Food Microbiology,* 2008. 25(1): p. 162–168.

[12] Painter J. A., et al., Attribution of Foodborne Illnesses, Hospitalizations, and Deaths to Food Commodities by using Outbreak Data, United States, 1998–2008. *Emerg Infect Dis.*, 2013.19(3): p. 407-415.

[13] Parish, M. E., et al., Methods to reduce/eliminate pathogens from fresh and fresh-cut produce. *Comprehensive Reviews in Food Science and Food Safety*, 2003. 2: p. 161–173.

[14] Mahajan P. V., et al., Postharvest treatments of fresh produce. *Phil. Trans. R. Soc.,* 2014. 372(2017): p. 1-19.

[15] Ioannou, I., and Ghoul, M., Prevention of enzymatic browning in fruit and vegetables *European Scientific Journal*, 2013. 9(30): p. 1857–7881.

[16] Artés, F., and Allende, A., Minimal fresh processing of vegetables, fruits and juices. In S. Da-Wen (Ed.), *Emerging technologies for food processing,* 2005. p. 677–716, London Academic Press.

[17] Rico, D., et al., Extending and measuring the quality of fresh-cut fruit and vegetables: A review. *Trends in Food Science & Technology*, 2007. 18(7): p. 373–386.

[18] Baur, S., et al., Effect of temperature and chlorination of pre-washing water on shelf-life and physiological properties of ready-to-use Iceberg lettuce. *Innovative Food Science & Emerging Technologies*, 2005. 6(2): p.171–182.

[19] Beltrán, D., et al., Effect of different sanitizers on microbial and sensory quality of fresh-cut potato strips stored under modified atmosphere or vacuum packaging. *Postharvest Biology and Technology,* 2005. 37(1): p. 37–46.

[20] Casteel, M. J., et al., Chlorine disinfection of produce to inactivate hepatitis A virus and coliphage MS2. *International Journal of Food Microbiology,* 2008. 125(3): p. 267–273.

[21] Hua, G. H., and Reckhow, D. A., Comparison of disinfection byproduct formation from chlorine and alternative disinfectants. *Water Research*, 2007. 41(8): p. 1667–1678.

[22] Keskinen, L. A., et al., Efficacy of chlorine, acidic electrolyzed water and aqueous chlorine dioxide solutions to decontaminate Escherichia coli O157:H7 from lettuce leaves. *International Journal of Food Microbiology,* 2009. 132(2–3): p. 134–140.

[23] Francis, G. A., and O'Beirne, D., Variation among strains of Listeria monocytogenes: Differences in survival on packaged vegetables and in response to heat and acid conditions. *Karlsruhe Nutrition Congress on Food Safety*, 2002. (p. 687–694). Germany Elsevier Sci Ltd.

[24] Fu, Y. C., et al., Effects of aqueous chlorine dioxide treatment on polyphenol oxidases from Golden Delicious apple. *LWT-Food Science and Technology*, 2007. 40(8): p. 1362–1368.

[25] Gómez-López, V. M., et al., Shelf-life extension of minimally processed carrots by gaseous chlorine dioxide. *International Journal of Food Microbiology*, 2007. 116(2): p. 221–227.

[26] Du, J., Han, Y., and Linton, R. H., Efficacy of chlorine dioxide gas in reducing Escherichia coli O157:H7 on apple surfaces. *Food Microbiology,* 2003. 20(5): p. 583–591.

[27] Gómez-López, V. M., et al., Shelf-life of minimally processed lettuce and cabbage treated with gaseous chlorine dioxide and cysteine. *International Journal of Food Microbiology*, 2008. 121(1): p. 74–83.

[28] Han, Y., et al., Inactivation of Escherichia coli O157:H7 on surface-uninjured and injured green pepper (Capsicum annuum L.) by chlorine dioxide gas as demonstrated by confocal laser scanning microscopy. *Food Microbiology*, 2000. 17(6): p. 643–655.

[29] Mahmoud, B. S. M., et al., Inactivation kinetics of inoculated Escherichia coli O157:H7, Listeria monocytogenes and Salmonella enterica on strawberries by chlorine dioxide gas. *Food Microbiology*, 2007. 24(7–8): p. 736–744.

[30] Singh, N., et al., Efficacy of chlorine dioxide, ozone, and thyme essential oil or a sequentialwashing in killing Escherichia coli O157:

H7 on lettuce and baby carrots. *LWT-Food Science and Technology*, 2002. 35(8): p. 720–729.

[31] Beuchat, L. R., *Surface decontamination of fruits and vegetables eaten raw: A review.* Food Safety Unit, World Health Organisation, 1998. WHO/FSF/FOS/98.2, 42.

[32] Ayala-Zavala, J., and Gustavo, G. A., Use of additives to preserve the quality of fresh-cut fruits and vegetables. In R. Soliva-Fortuny (Ed.), *Advances in fresh-cut fruits and vegetables processing*, 2010. (p. 231–254). CRC Press.

[33] Aase, B., et al., Occurrence of and a possible mechanism for resistance to a quaternary ammonium compound in Listeria monocytogenes. *International Journal of Food Microbiology*, 2000. 62(1–2): p. 57–63.

[34] To, M. S., et al., Post adaptational resistance to benzalkonium chloride and subsequent physicochemical modifications of Listeria monocytogenes. *Applied and Environmental Microbiology*, 2002. 68(11): p. 5258–5264.

[35] Velazquez, L. D., et al., Evaluation of chlorine, benzalkonium chloride and lactic acid as sanitizers for reducing Escherichia coli O157:H7 and Yersinia enterocolitica on fresh vegetables. *Food Control,* 2009. 20(3): p. 262–268.

[36] Ölmez, H., and Kretzschmar, U., Potential alternative disinfection methods for organic fresh-cut industry for minimizing water consumption and environmental impact. *LWT-Food Science and Technology*, 2009. 42(3): p. 686–693.

[37] Akbas, M. Y., and Olmez, H., Effectiveness of organic acid, ozonated water and chlorine dippings on microbial reduction and storage quality of fresh-cut Iceberg lettuce. *Journal of the Science of Food and Agriculture*, 2007. 87(14): p. 2609–2616.

[38] Alexandre, E. M. C., et al., Assessment of the impact of hydrogen peroxide solutions on microbial loads and quality factors of red bell peppers, strawberries and watercress. *Food Control*, 2012. 27(2): p. 362–368.

[39] Hwang, E. S., et al., Ozone and hydrogen peroxyacetic acid treatment to reduce or remove EBDCs and ETU residues in a solution. *Journal of Agricultural and Food Chemistry*, 2001. 49(11): p. 5689–5694.

[40] Sapers, G. M., Washing and sanitizing treatments for fruits and vegetables. *Microbiology of Fruits and Vegetables*, 2006. p. 375–400.

[41] Martin-Diana, A. B., et al., Calcium lactate washing treatments for salad-cut Iceberg lettuce: Effect of temperature and concentration on quality retention parameters. *Food Research International*, 2005. 38(7): p. 729–740.

[42] Kim, J. G., et al., Applications of ozone for enhancing the microbiological safety and quality of foods: A review. *Journal of Food Protection*, 2009. 62(9): p. 1071-1087.

[43] Grass, M. L., et al., Calcium fortification of vegetables by vacuum impregnation interactions with cellular matrix. *Journal of Food Engineering,* 2003. 56: p. 279-284.

[44] Beltrán, D., et al., Ozonated water extends the shelf life of fresh-cut lettuce. *Journal of Agricultural and Food Chemistry*, 2005. 53(14): p. 5654–5663.

[45] Skog, L. J., and Chu, C. L., Effect of ozone on qualities of fruits and vegetables in cold storage. *Canadian Journal of Plant Science*, 2001. 81(4): p. 773-778.

[46] White, G. C., Ozone. In: *Handbook of Chlorination and Alternative Disinfectants.* Third Edition. Van Nostrand Reinhold, New York. 1992. p. 1046-1110.

[47] Guzel-Seydim, Z. B., et al., Use of ozone in the food industry. LWT *Food Science and Technology*, 2004. 37: p. 453-460.

[48] Smilanick, J. L., et al., Postharvest use of ozone on fresh fruit. *Perishables Handling Quarterly Issue*, 1999. 99: p. 10-14.

[49] Kim, C., et al., Roles of oxidation–reduction potential in electrolyzed oxidizing and chemically modified water for the inactivation of food-related pathogens. *Journal of Food Protection*, 2000. 63: p. 19–24.

[50] Bari, M. L., et al., Effectiveness of electrolyzed acidic water in killing Escherichia colli O157:H7, Salmonella enteridis, and Listeria

monocytogenes on the surfaces of tomatoes. *Journal of food protection,* 2003. 66: p. 542-548.

[51] Koseki, S., et al., Decontamination of lettuce using acidic electrolyzed water. *Journal of Food Protection*, 2001. 64: p. 652-658.

[52] Polanska, M., *Development of a new method for disinfection of drinking water.* M.Sc. dissertation. Katholieke Hogeschool Sint-Lieven, Gent, Belgium and Wroclaw University of Technology, Wroclaw, Poland, 2001. p. 1–156.

[53] Martin-Diana, A. et al., Extending and Measuring the Quality of Fresh-Cut Fruit and Vegetables: a Review. *Trends in Food Technology*, 2007. 18 (7): p. 373-386.

[54] Roller, S., *Natural antimicrobials for the minimal processing of foods.* Roller, S. (ed.). 2003. Woodhead Publishing. Cambridge, UK.

[55] Martín-Diana, A. B., et al., Whey permeate as bio-preservative for shelf-life maintenance of fresh-cut vegetables. *Innovative Food Science &Technologies*, 2006. 7: p. 112-123.

[56] Bari, M. L., et al., Combined efficacy of nisin and pediocin with sodium lactate, citric acid, phytic acid, and potassium sorbate and EDTA in reducing the Listeria monocytogenes population of inoculated fresh-cut produce. *Journal of Food Protection*, 2005. 68: p. 1381–1387.

[57] Özel, A., et al., Purification and characterisation of a polyphenol oxidase from Boletus erythropus and investigation of its catalytic efficiency in selected organic solvents. *Food Chemistry*, 2010. 119(3): p. 1044-1049.

[58] Lavelli, V., and Caronni, P., Polyphenol oxidase activity and implications on the quality of intermediate moisture and dried apples. *European Food Research and Technology*, 2010. 231(1): p. 93-100.

[59] Ingraham, L. L., Reaction-Inactivation of Polyphenol Oxydase: Catachol and Oxygen Dependence. *Journal of the American Chemical Society*, 1995. 77(10): p. 2875-2876.

[60] Kregiel, K., "Health Safety of Soft Drinks: Contents, Containers, and Microorganisms," *BioMed Research International*, 2015. 2015: p. 1-15.

[61] *Canadean Beverage Categories Definitions, 2008*, http://www.liquidforecasts.com/CanadeanDefinitions.pdf.

[62] Sohaib, M., et al., Postharvest intervention technologies for safety enhancement of meat and meat based products; a critical review, *J Food Sci Technol*, 2016. 53(1): p.19–30.

[63] Lawrie, R. A., and Ledward, D. A., *Lawrie's Meat Science, 7th edn. 2006.* Woodhead Publishing Limited, Cambridge.

[64] Loeffler, M., et al., Electrostatic interactions of cationic lauric arginate with anionic polysaccharides affect antimicrobial activity against spoilage yeasts. *J Appl Microbiol,* 2014. 117: p. 28–39.

[65] Zhou, G. H., et al., Preservation technologies for fresh meat: A review. *Meat Sci*, 2010. 86: p. 119–128.

[66] Mahony, J., et al, Bacteriophages as biocontrol agents of food pathogens. *Current Opinion in Biotechnology*, 2010, 22(2): p. 127-163.

[67] Hugas, M., et al., New mild technologies in meat processing: high pressure as a model technology. *Meat Sci,* 2002. 62: p. 359–371.

[68] Sofos, J. N., and Busta, F. F. Chemical food preservatives. In: Russell AD, Hugo WB, Aycliffe GAJ (eds) *Principles and Practice of Disinfection, Preservation and Sterilization,* 2nd edn., 2001. Blackwell Scientific, Oxford.

[69] Buttriss, J. L., Food reformulation: the challenges to the food industry, Conference on 'Future food and health' Symposium III: Food innovation and health, *Proceedings of the Nutrition Society*, 2013. 72: p. 61–69.

[70] He and MacGregor, A comprehensive review on salt and health and current experience of worldwide salt reduction programmes. *Journal of Human Hypertension*, 2009. 23: p. 363–384.

[71] *European Commission Directorate-General Health and Consumers, Survey on members states' implementation of the EU salt reduction framework, 2013.* http://ec.europa.eu/health/nutrition_physical_activity/docs/salt_report1_en.pdf 2013.

[72] World Health Organization Regional Office for Europe, *Mapping salt reduction initiatives in the WHO European Region*, 2013.

[73] http://www.euro.who.int/__data/assets/pdf_file/0009/186462/Mapping-salt-reduction

[74] Kloss, L., et al., Sodium intake and its reduction by food reformulation in the European Union — A review, *NFS Journal*, 2015. 1: p. 9–19.

[75] Powles, J., et al., Global, regional and national sodium intakes in 1990 and 2010: a systematic analysis of 24 h urinary sodium excretion and dietary surveys worldwide, *BMJ Open*, 2013. 3: p. 1–20.

[76] Dötsch, M., et al., Strategies to reduce sodium consumption: a food industry perspective, *Crit. Rev. Food Sci. Nutr.*, 2009. 49: p. 841–851.

[77] Anderson, C.A., et al., Dietary sources of sodium in China, Japan, the United Kingdom, and the United States, women and men aged 40 to 59 years: the INTERMAP study, *J. Am. Diet. Assoc.*, 2010. 110: p. 736–745.

[78] Wilson, R., et al., Evaluation of technological approaches to salt reduction. *Leatherhead Food Research*. 2012.

[79] https://www.fdf.org.uk/resources/salt_reduction_2012.pdf 2012.

[80] Belz, M. C. E., et al., The impact of salt reduction in bread: a review, *Crit. Rev. Food Sci. Nutr.*, 2012. 52: p. 514–524.

[81] Hutton, T., Sodium technological functions of salt in the manufacturing of food and drink products, *BFJ,* 2002. 104(2): p. 126–152.

[82] Tournay, R., Salt and satiety, *Eur. Baker*, 2009. p. 16–18.

[83] Charlton, K. E., et al., Partial replacement of NaCl can be achieved with potassium, magnesium and calcium salts in brown bread, *Int. J. Food Sci. Nutr.,*2007. 58(7): p. 508–521.

[84] Minasian, S., Permeate can help you reduce sodium while improving flavor, *Dairy Pipeline*, 2011. 23(2): p. 3–8.

[85] Desmond, E., Reducing salt: a challenge for the meat industry, *Meat Sci.*, 2006. 74: p. 188–196.

[86] Devlieghere, F., et al., Reducing salt intake from meat products by combined use of lactate and diacetate salts without affecting microbial stability, *Int. J. Food Sci. Technol.*, 2009. 44: p. 337–341.

[87] Aliño, M., et al., J. M. Barat, Influence of sodium replacement on physicochemical properties of dry-cured loin, *Meat Sci.,* 2009. 83: p. 423–430.

[88] Gimeno, O. et al., Calcium ascorbate as a potential partial substitute for NaCl in dry fermented sausages: effect on color, texture and hygienic quality at different concentrations, *Meat Sci.*, 2001. 57: p. 23–29.

[89] Barat, M., et al., Partial replacement of sodium in meat and fish products by using magnesium salts. A review, *Plant Soil*, 2013. 368: p.179–188.

[90] Ruusunen, M., et al., The effect of fat content and flavour enhancers on the perceived saltiness of cooked bologna-type sausages, *J. Muscle Foods*, 2001. 12: p. 107–120.

[91] Searby, L., Pass the salt, *International Food Ingredients*, 2006. p. 6–8.

[92] Johnson, M. E., et al., Reduction of sodiumand fat levels in natural and processed cheeses: scientific and technological aspects, *Compr. Rev. Food Sci. Food Saf.*, 2009. 8 (3): p. 252–268.

[93] Phan, V. A., *In vivo* sodium release related to salty perception during eatingmodel cheeses of different texture, *Int. Dairy J.*, 2008. 18: p. 956–963.

[94] Saint–Eve, C. et al., Reducing salt and fat content: impact of composition, texture and cognitive interactions on the perception of flavoured model cheeses, *Food Chem.,* 2009. 116(1): p. 167–175.

[95] Fitzgerald, E., and Buckley, J., Effect of total and partial substitution of sodium chloride on the quality of cheddar cheese, *J. Dairy Sci.,* 1985. 68: p. 3127–3134.

[96] Lesage, L., et al., Sodium chloride andmagnesium chloride affected by ripening of Camembert cheese, *J. Food Sci.*, 1993. 58: p. 1303–1306.

[97] Gupta, S. K., et al., Effect of emulsifier salts on textural and flavor properties of processed cheeses, *J. Dairy Sci.*, 1984. 67: p. 764–778.

[98] Burgos, K. et al., *Reformulation guide for small to medium sized companies, 2016*. Leatherhead Food Research.

[99] Yao, C. K. et al., "Dietary sorbitol and mannitol: food content and distinct absorption patterns between healthy individuals and patients with irritable bowel syndrome," *Journal of Human Nutrition and Dietetics*, 2014. 27(2): p. 263–275.

[100] Tuck, C. J., et al., "Fermentable oligosaccharides, disaccharides, monosaccharides and polyols: role in irritable bowel syndrome," *Expert Review of Gastroenterology and Hepatology*, 2014. 8(7): p. 819–834.

[101] *EFSA Scientific Opinion on Neotame, September 2007.*

[102] *EFSA Opinion on Advantame, July 2013.*

[103] *WHO, Healthy diet, 2015, Fact sheet N°394*, Updated September 2015, http://www.who.int/mediacentre/factsheets/fs394/en/

[104] AHA, Dietary guidelines for healthy Americans. *Circulation,* 1996. 94: p. 1795-1800.

[105] *USDHHS, "The Surgeon General's Report on Nutrition and Health. Publ. No. 88-50210. 1988.* U.S. Govt. Print. Office, Washington, D. C.

[106] EUFIC, *Food innovation and reformulation for a healthier Europe: A challenging mission,* 2010.

[107] Akoh, C. C., Fat replacers, A publication of the Institute of food technologists 'Expert panel on food safety and nutrition, *Food Technology,* 1998. 52 (3).

[108] Miele, N. A., *Calorie reduction in food: Sensory performance of a new sweetener and fat replacer optimization*, 2013.

[109] Akoh, C. C. and Swanson, B. G., Synthesis of carbohydrate fatty acid polyesters. In *"Carbohydrate Polyesters as Fat Substitutes,"* ed. C.C. 1994 (p. 9-35), Marcel Dekker, Inc., N.Y.

[110] Akoh, C. C. and Swanson, B.G., Optimized synthesis of sucrose polyesters: Comparison of physical properties of sucrose polyesters, raffinose polyesters and salad oils. *J. Food Sci.,* 1990. 55: p. 236-243.

[111] Rizzi, G. P., and Taylor, H. M., A solvent-free synthesis of sucrose polyesters. *J. Am. Oil Chem. Soc.*, 1978. 55: p. 398-401.

[112] Shieh, C. J., et al., Optimization of sucrose polyester synthesis using response surface methodology. *J. Food Sci.*, 1996. 61: p. 97-100.

[113] Osipow, L., et al., Methods of preparation of fatty acid esters of sucrose. *Ind. Eng. Chem.*, 1956. 48: p. 1459-1462.

[114] Harrigan, K. A., and Breene, W. M., Fat substitutes: Sucrose polyesters and other synthetic oils. In "*Low- Calorie Foods Handbook*," ed. A. M. Altschul, 1993. Marcel Dekker, Inc., N.Y.

[115] Kabara, J. J., Fatty acids and derivatives as antimicrobial agents: A review. In "*Pharmacological Effect of Lipids,*" ed. J.J. Kabara, 1998. p. 1-14, *Amer. Oil Chem. Soc.,* Champaign, Ill.

[116] Marshall, D. L., and Bullerman, L. B., Antimicrobial properties of sucrose fatty acid esters. In *"Carbohydrate Polyesters as Fat Substitutes,"* ed. C. C. Akoh and B. G. Swanson, 1994. p. 149-167, Marcel Dekker, Inc., N.Y.

[117] Unilever NV., *Preparation of polyol fatty acid polyesters.* Netherlands, 1988. Patent 8,601,904.

[118] Akoh, C. C., Structured lipids - enzymatic approach. *INFORM*. 1995. 6: p. 1055-1061.

[119] Heird, W. C., et al., Structured lipids and their use in clinical nutrition. *Am. J. Clin. Nutr.,* 1986. 43: p. 320-324.

[120] Kennedy, J. P., Structured lipids: Fats of the future. *Food Technol.*, 1991. 45(11): p. 76- 83.

[121] Babayan, V. K., Specialty lipids and their biofunctionality. *Lipids,* 1987. 22: p. 417-420.

[122] Bach, A. C., et al., The usefulness of dietary medium-chain triglycerides in body weight control: Fact or fancy? *J. Lipid Res.*, 1996. 37: p. 708-726.

[123] Megremis, C. J., Medium-chain triacylglycerols: A nonconventional fat. *Food Technol.*, 1991. 45(2): p. 108-114.

[124] Kosmark, R., Salatrim: Properties and applications. *Food Technol.*, 1996. 50(4): p. 98-101.

[125] Hamm, D. J., *Low calorie edible oil substitutes.* U.S. Patent, 1985. 4,508,746.

[126] Marcus, J. B., Lipids Basics: Fats and Oils in Foods and Health: Healthy Lipid Choices, Roles and Applications in Nutrition, Food

Science and the Culinary Arts. Culinary Nutrition, *The Science and Practice of Healthy Cooking* 2013, 6: p. 231–277.

[127] Giese, J., Fats, oils, and fat replacers. *Food Technol.*, 1996. 50(4): p. 78-84.

[128] Allen, L., et al., *Guidelines on food fortification with micronutrients, 2006, World Health Organization and Food and Agriculture Organization of the United Nations.*

[129] Hurrell, R.F., How to ensure adequate iron absorption from iron-fortified food. *Nutrition Reviews*, 2002. 60 (7 Pt 2): p. S7–S15.

[130] Hurrell, R. F., Iron. In: Hurrell RF, ed. *The mineral fortification of food. Leatherhead,* Surrey, Leatherhead Publishing, 1999. p. 54–93.

[131] Swain, J. H., et al., Bioavailability of elemental iron powders to rats is less than bakery-grade ferrous sulfate and predicted by iron solubility and particle surface area. *Journal of Nutrition*, 2003. 133: p. 3546–3552.

[132] Cori, H., *Personal communication*, 2004.

[133] Dary O., and Mora, J. O., Food fortification to reduce vitamin A deficiency: International Vitamin A Consultative Group recommendations. *Journal of Nutrition*, 2002. 132 (9): p. 2927S–2933S.

[134] Rosado, J. L., et al., Development, production, and quality control of nutritionnal supplements for a national supplementation programme in Mexico. *Food and Nutrition Bulletin*, 2000, 21: p. 30–34.

[135] Tartanac, F., Incaparina and other Incaparina-based foods: Experience of INCAP in Central America. *Food and Nutrition* Bulletin, 2000. 21: p. 49–54.

[136] Lopez de Romana, D., Experience with complementary feeding in the FONCODES project. *Food and Nutrition Bulletin*, 2000. 21: p. 43–48.

[137] van Stuijvenberg, M. E., et al., Long-term evaluation of a micronutrient-fortified biscuit used for addressing micronutrient deficiencies in primary school children. *Public Health Nutrition*, 2001. 4: p. 1201–1209.

[138] Chavasit, V., and Tontisirin, K., Triple fortification: instant noodles in Thailand. In: *Micronutrient Initiative*, ed. *Food fortification to end micronutrient malnutrition.* State of the Art. Ottawa, Micronutrient Initiative, 1998. p. 72–76.

[139] Bynum, D., Fortification of dairy products with micronutrients to end malnutrition. In: *Micronutrient Initiative*, ed. *Food fortification to end micronutrient malnutrition*. State of the Art. Ottawa, Micronutrient Initiative, 1998. P. 38–42.

[140] Delange, F., and Hetzel, B. S., The iodine deficiency disorders. In: Hennemann G, DeGroot L, eds. *The thyroid and its diseases*. 2003. MA, Endocrine Education, Inc.

[141] Gerasimov, G., et al. Bread iodization for iodine deficient regions of Russia and other newly independant states. *IDD Newsletter*, 1997. 13: p.12–13.

[142] Phillips, D. I., Iodine, milk, and the elimination of endemic goiter in Britain: the story of an accidental public health triumph. Journal of *Epidemiology and Community Health*, 1997. 51: p. 391–393.

[143] Suwanik, R., et al., Simple technology provides effective IDD control at the village level in Thailand. *IDD Newsletter,* 1989. 5: p. 1–6.

[144] Đorđević, et al., Trends in Encapsulation Technologies for Delivery of Food Bioactive Compounds, *Food Eng Rev,* 2015. 7: p. 452–490.

[145] Ghosh, S. K., Functional Coatings and Microencapsulation: a General Perspective, in *Functional Coatings*, 2006. Weinheim: Wiley-VCH.

[146] Kuang, S. S., et al., Microencapsulation as a Tool for Incorporating Bioactive Ingredients into Food, *Critical Reviews in Food Science and Nutrition,* 2010. 50(10): p. 951-968.

[147] Gibbs, B. F., et al., Encapsulation in the food industry: A review. *Internatioanl Journal of Food Sciences and Nutrition*, 1999. 50: p. 213–224.

[148] Gouin, S., Micro-encapsulation: Industrial appraisal of existing technologies and trends. *Trends in Food Science and Technology*, 2004. 15: p. 330–347.

[149] Fang, Z., and Bhandari, B., Encapsulation of polyphenols-a review, *Trends Food Sci.*, 2010. 21: p. 510-523.

[150] Young, S. L., et al., Microencapsulation properties of whey proteins. 1. Microencapsulation of anhydrous milk fat. *Journal of Dairy Science,* 1993. 76: p. 2868–2877.

[151] Rosenberg, M., and Sheu, T. Y., Microencapsulation of volatiles by spray-drying in whey protein-based wall systems. *International Dairy Journal,* 1996. 6: p. 273–284.

[152] Baranauskiene´, R., et al., Properties of oregano (Origanum vulgare L.), citronella (Cymbopogon nardus G.) and marjoram (Majorana hortensis L.) flavors encapsulated into milk protein-based matrices. *Food Research International*, 2006. 39: p. 413–425.

[153] Hogan, S. A., et al., Microencapsulation properties of sodium caseinate. *Journal of Agricultural and Food Chemistry*, 2001. 49: p. 1934–1938.

[154] Oneda, F., and Re´, M. I., The effect of formulation variables on the dissolution and physical properties of spray-dried microspheres containing organic salts. *Powder Technology*, 2003. 130: p. 377–384.

[155] Shu, B., et al., Study on microencapsulation of lycopene by spray-drying. *Journal of Food Engineering*, 2006. 76: p. 664–669.

[156] Drusch, S., et al., Physicochemical characterization and oxidative stability of fish oil encapsulated in an amorphous matrix containing trehalose. *Food Research International*, 2006. 39: p. 807–815.

[157] Watanabe, Y., et al., Oxidation of 6-O-arachidonoyl -ascorbate microencapsulated with a polysaccharide by spray-drying. *Lebensmittel-Wissenschaft und-Technologie*, 2004. 37: p. 395–400.

[158] Krishnan, S., et al., Microencapsulation of cardamom oleoresin: Evaluation of blends of gum arabic, maltodextrin and a modified starch as wall materials. *Carbohydrate Polymers*, 2005. 61: p. 95–102.

[159] Soottitantawat, A., et al., Influence of emulsion and powder size on the stability of encapsulated D-limonene by spray drying. Innovative *Food Science and Emerging Technologies*, 2005. 6: p. 107–114.

[160] Soottitantawat, A., et al., Microencapsulation of L-menthol by spray drying and its release characteristics. *Innovative Food Science and Emerging Technologies*, 2005. 6: p. 163–170.

[161] Teixeira, M. I., et al., Characterization of short chain fatty acid microcapsules produced by spray drying. *Materials Science and Engineering,* 2004. 24: p. 653–658.

[162] Borgogna et al., Food microencapsulation of bioactive compounds: Rheological and thermal characterization of non-conventional gelling system, *Food Chemistry*, 2010. 122(2): p. 416-423.

[163] Peanparkdee, M., Iwamoto, S., Yamauchi R., Microencapsulation: a review of applications in the food and pharmaceutical industries, *Reviews in Agricultural Science*, 2016. 4: p. 56-65.

[164] Burin, V. M., et al., Anthocyanins: optimisation of extraction from Cabernet Sauvignon grapes, microcapsulation and stability in soft drink. *Int. J. Food Sci. Technol.*, 2011. 46: p. 186-193.

[165] Giusti, M. M., and Wrolstad, R. E., Acylated anthocyanins from edible sources and their applications in food systems. *Biochem. Eng. J.,* 2003. 14: p. 217-225.

[166] Aditya, N., et al., Curcumin and catechin co-loaded water-in-oil-in-water emulsion and its beverage application. *J. Funct. Foods*, 2015. 15: p. 35-43.

[167] Green, R. J., et al., Common tea formulations modulate *in vitro* digestive recovery of green tea catechins. *Mol. Nutr. Food Res.*, 2007, 51: p. 1152-1162.

[168] Kausadikar, S., et al., Microencapsulation of lemon oil by spray drying and its application in flavour tea. *Adv. Appl. Sci. Res.*, 2015. 6: p. 69-78.

[169] Rocha, G. A., et al., Microencapsulation of lycopene by spray drying: characterization, stability and application of microcapsules. *Food Bioprod. Process.*, 2012. 90: p. 37-42.

[170] O'Brien, C., et al., Effect of varying the microencapsulation process on the functionality of hydrogenated vegetable fat in shortdough biscuits. *Food Res. Int.,* 2003. 36: p. 215-221.

[171] Shah, N., and Ravula, R., Microencapsulation of probiotic bacteria and their survival in frozen fermented dairy desserts. *Aust. J. Dairy Technol.,* 2000. 55: p. 139+

[172] Kailasapathy, K., and Rybka, S., L. acidophilus and Bifidobacterium spp.-their therapeutic potential and survival in yogurt. *Austr. J. Dairy Technol.,* 1997. 52: p. 28+

[173] Lücke, F. K, Utilization of microbes to process and preserve meat. *Meat Sci.*, 2000. 56: p. 105-115.

[174] Audet, P., et al., Immobilized growing lactic acid bacteria with κ-carrageenan–locust bean gum gel. *Appl. Microbiol. Biotechnol.,* 1988. 29: p. 11-18.

[175] Muthukumarasamy, P., and Holley, R. A., Microbiological and sensory quality of dry fermented sausages containing alginate-microencapsulated Lactobacillus reuteri. *Int. J. Food Microbiol.*, 2006. 111: p. 164-169.

[176] Jiménez-Martín, E., et al., Enrichment of Chicken Nuggets with Microencapsulated Omega-3 Fish Oil: Effect of Frozen Storage Time on Oxidative Stability and Sensory Quality. *Food Bioprocess Technol.*, 2016. 9: p. 285-297.

[177] Comunian. A., et al., Effect of Incorporating Free or Encapsulated Ascorbic Acid in Chicken Frankfurters on Physicochemical and Sensory Stability. *J. Food Sci. Eng.*, 2014. p. 167-175.

[178] Anjani. K., et al., Microencapsulation of enzymes for potential application in acceleration of cheese ripening. *Int. Dairy J.,* 2007. 17: p. 79-86.

[179] Champagne, C. P., et al, Effects of storage conditions, microencapsulation and inclusion in chocolate particles on the stability of probiotic bacteria in ice cream. *Int. Dairy J.,* 2015. 47: p. 109-117.

[180] Ortigoza, et al, Advances in the Functional Characterization and Extraction Processes of Dietary Fiber, *Food Eng Rev*, 2016. 8: p. 251–271.

[181] *Codex Alimentarius Commission (CAC) JOINT FAO/WHO Food Standards Programme, Codex Alimentarius Commission, Rome 2009, Thirty second Session.*

[182] Angioloni. A., and Collar, C., Small and large deformation viscoelastic behaviour of selected fibre blends with gelling properties. *Food Hydrocoll*, 2009. 23: p. 742–748.

[183] Maran, J. P., and Priya, B., Ultrasound-assisted extraction of pectin from sisal waste. *Carbohydr Polym*, 2015. 115: p. 732–738.

[184] Wang, S., et al., Optimization of pectin extraction assisted by microwave from apple pomace using response surface methodology. *J Food Eng,* 2007. 78: p. 693–700.

[185] Wang, J., et al., Optimisation of ultrasound- assisted enzymatic extraction of arabinoxylan from wheat bran. *Food Chem*, 2014. 150: p. 482–488.

[186] Xu, Y., et al., Effects of ultrasound and/or heating on the extraction of pectin from grapefruit peel. *J Food Eng*, 2014. 126: p. 72–81.

[187] De Vries. S., et al., Improving digestive utilization of fiber-rich feedstuffs in pigs and poultry by processing and enzyme technologies: a review. *Anim Feed Sci Technol*, 2012. 178: p. 123–138.

[188] Guo. X., et al., Extraction of pectin from navel orange peel assisted by ultra-high pressure, microwave or traditional heating: a comparison. *Carbohydr Polym,* 2012. 88: p. 441–448.

[189] Liu, Y., et al., Water-based extraction of pectin from flavedo and albedo of orange peels. *Chem Eng J,* 2006. 120: p. 203–209.

[190] Wang, X., et al., Pectin extracted from apple pomace and citrus peel by subcritical water. *Food Hydrocoll*, 2014. 38: p. 129–137.

[191] Izydorczyk, M. S., and Biliaderis, C. G., Cereal arabinoxylans: advances in structure and physicochemical properties. *Carbohydr Polym*, 1995. 28: p. 33–48.

[192] Jacquemin, L., et al., Evaluation of the technical and environmental performances of extraction and purification processes of arabinoxylans from wheat straw and bran. *Process Biochem*, 2012. 47: p. 373–380.

[193] Zhang, Z., et al., Extraction and modification technology of arabinoxylans from cereal by-products: a critical review. *Food Res Int*, 2014. 65(C): p. 423-436.

[194] Chanliaud. E., et al., Alkaline extraction and characterisation of heteroxylans from maize bran. *J Cereal Sci*, 1995. 21: p. 195–203.

[195] Mueller. K., et al., Simplified fractionation process for linseed meal by alkaline extraction: functional properties of protein and fibre fractions. *J Food Eng*, 2010. 99: p. 49–54.

[196] Wikiera. A., et al., Multicatalytic enzyme preparations as effective alternative to acid in pectin extraction. *Food Hydrocoll*, 2015. 44: p. 156–161.

[197] Xiong. X., et al., Effects of alkali treatment and subsequent acidic extraction on the properties of soybean soluble polysaccharides. *Food Bioprod Process*., 2015. 94: p. 239-247.

[198] Chen, L., and Anderson, W., Soluble and insoluble plan fiber in selected cereals and vegetables. *Am J Clin Nutr*, 1981. 34: p. 1077–1082.

[199] Jolie, R. P., et al., Pectin conversions under high pressure: implications for the structurerelated quality characteristics of plant-based foods. *Trends Food Sci Technol*, 2012. 24: p. 103–118.

[200] Sarkar, N., et al., Bioethanol production from agricultural wastes: an overview. *Renew Energy*, 2012. 37: p. 19–27.

[201] Gamel, T. H., et al., Enzymatic extraction of beta-glucan from oat bran cereals and oat crackers and optimization of viscosity measurement. *J Cereal Sci*, 2014. 59: p. 33–40.

[202] Villanueva-Suarez, M. J., et al., Sequential extraction of poly-saccharides from enzymatically hydrolyzed okara byproduct: physicochemical properties and *in vitro* fermentability. *Food Chem*, 2013. 141: p. 1114–1119.

[203] Meyer, A. S., et al., Enzymatic solubilisation of a pectinaceous dietary fiber fraction from potato pulp: optimization of the fiber extraction process. *Biochem Eng J,* 2009. 43: p. 106–112.

[204] Jeong, H. S., et al., Optimization of enzymatic hydrolysis conditions for extraction of pectin from rapeseed cake (Brassica napus L.) using commercial enzymes. *Food Chem*, 2014. 157: p. 332–338.

[205] Dominiak, M., et al., Application of enzymes for efficient extraction, modification, and development of functional properties of lime pectin. *Food Hydrocoll*, 2014. 40: p. 273–282.

[206] Qi, B., et al., Extract dietary fiber from the soy pods by chemistry-enzymatic methods. *Procedia Eng*, 2011. 15: p. 4862–4873.

[207] Lusas, E. W., et al., Producing edible oil and protein from oilseeds by aqueous processing, *Oil Mill Gazetteer*, 1982. 86: p. 28+

[208] Considine, D. M., and Considine, G. D., *Foods and Food Production Encyclopedia*, 1982. Van Nostrand Reinhold Co. Inc., New York.

[209] Visser, A., and Thomas, A., Review: Soya protein products—Their processing, functionality, and application aspects, *Food Rev*. 1987. 3: p. 1+.

[210] Berk, Z., *Technology of Production of Edible Flours and Protein Products from Soybeans*, FAO Agricultural Services Bulletin.

[211] Oreopoulou, V., and Tzia, C., Utilization of Plant By-Products for the Recovery of Proteins, Dietary Fibers, Antioxidants, and Colorants, 11: p. 209-233 In: *Utilization of By-Products and Treatment of Waste in the Food Industry,* Editors: Oreopoulou, Vasso, Russ, Winfried, 2007, Springer.

[212] Milligan, E. D., et al., Processing for production of edible soya flour, *JAOCS*, 1981. 58: p. 331.

[213] Tandy, D. C., Oilseed extraction, In: *Introduction to Fats and Oils Technology,* Wan, P. J., ed., 1991. p. 59–84, AOCS, Champaign.

[214] Bernardini, E., 1983, *Oilseeeds, Oils and Fats*, B. E. Oil Publishing House, Rome, pp. 361–363, 483–493.

[215] Mustakas, G. C., et al., Flash desolventizing defatted soybean meals washed with aqueous alcohols to yield a high-protein product, *JAOCS,* 1962. 39: p. 222+.

[216] Waggle, D. H., and Kolar, C.W., Type of soy protein products, In: *Soy Protein and Human Nutrition*, Wilcke, H. L., et al., eds., 1979. p. 19–52. Academic Press Inc., New York.

[217] Kosseoglu, S. S., and Rhee, K. C., New technologies in processing protein products, *Presented at the 85th AOCS Annual Meeting*, 1994. Atlanta, GA, USA.

[218] Lawhon, J. T., and Lusas, E. W., New techniques in membrane processing of oilseeds, *Food Technol.* 1984. 38: p. 97+.

[219] Liadakis, G., et al., Protein isolation from tomato seed meal, extraction optimisation, *J. Food Sci.,* 1995. 60: p. 477+.

[220] Park, S. H., and Bean, S. R., Investigation and optimization of the factors influencing sorghum protein extraction, *J. Sci. Food Agric.*, 2003 51: p. 7050+

[221] Tzia, C., Optimization methods, In: *Extraction Optimization in Food Engineering*, Tzia C., and Liadakis, G., eds., 2003. p. 137–180. Marcel Dekker Inc., New York.

[222] Mieth, G., and Kroll, J., Sonnenblumensamen und deren verarbeitungs produkte 2. Mitt. Gewinnung proteinangereicherter verarbeitungs produkte [Sunflower seeds and their processing products 2. Mid-level production of protein-enriched processed products], *Die Nahrung.*, 1984. 28: p. 907+.

[223] Yazicioglu, T., Studies on extraction of protein isolates from Turkish sunflowerseed and cottonseed extraction cakes and on their functional properties, *Fette Seifen Anstrichm.*, 1981. 83: p. 474+.

[224] Taha, F. S., Abbasy, M., El-Nockrashy, A. S., and Shoeb, Z. E., 1981, Countercurrent extraction-Isoelectric precipitation of sunflower seed protein isolates, *J. Sci. Food Agric.* 32:166.

[225] El-Tinay, A., et al., Protein and gossypol extractability from cottonseed flour, *J. Sci. Food Agric.*, 1980. 31: p. 38+.

[226] Wanasundara, P. K. J. P. D., and Shahidi, F., Optimization of hexametaphosphate assisted extraction of flaxseed proteins using response surface methodology, *J. Food Sci.,* 1966. 61: p. 604+.

[227] Eggers, L., *Some Biochemical and Electron Microscopic Studies of the Protein Present in Seeds Recovered from Tomato Cannery Waste,* Ph.D. Dissertation, 1974. The Ohio State Univ., Columbus, OH.

[228] Liadakis, G. N., et al., Isolation of tomato seed meal proteins with salt solutions, *J. Food Sci.*, 1998. 63: p. 450+.

[229] Tchorbanov, B., et al., Lipid-protein concentrations from tomato and paprika seeds, *Nahrung.,* 1986. 30: p. 408+.

[230] Bernardini, E., and Bernardini, M., Industrial apparatus for the production of protein superconcentrates from seed meals, *Trop. Sci.*, 1974. 16: p. 23+.

[231] Belter, P. A., et al., Bioseparations. *Downstream Processing for Biotechnology,* 1988. p. 221–236, Wiley, New York.

[232] Glatz, C. E., and Fisher, R. R., 1986, Modelling of precipitation phenomena in protein recovery, *ACS Symposium Series*, 314, pp. 109–120.

[233] Hamel, J. P., and Hunter, J. B., Modelling and applications of downstream processing, *ACS Symposium Series*, 1990. 419: p. 1–35.

[234] Wheelwright, S. M., *Protein Purification. Design and Scale up of Downstream Processing*, 1991. p. 88–91, Hanser Publishers, New York.

[235] Lawhon, J. T., Manak, L. J., Rhee, K. C., Rhee, K. S., and Lusas, E. W., 1981, Combining aqueous extraction and membrane isolation techniques to recover protein and oil from soybeans, *J. Food Sci.* 46:912.

[236] Schwass, D. E., and Finley, J. W., Heat and alkaline damage to proteins: racemization and lysinoalanine formation. *J. Agric. Food Chem.* 1984. 32: p. 1377–1382.

[237] Shen, L., et al., Studies on tea protein extraction using alkaline and enzyme methods. *Food Chem.*, 2008. 107: p. 929–938.

[238] Khan, M. A. A., et al., Effect of enzymatic fish-scrap protein hydrolysate on gelforming ability and denaturation of lizard fish Saurida wanieso surimi during frozen storage. *Fisheries Sci*, 2003. 69(6): p. 1271–1280.

[239] Morimura, S., et al., Development of an effective process for utilization of collagen from livestock and fish waste. *Process Biochem*, 2002. 37(12): p. 1403–1412.

[240] Batista, I., Recovery of proteins from fish waste products by alkaline extraction. *European Food Research Technol*, 1999. 210(2): p. 84–89.

[241] Hultin, H. O., and Kelleher, S. D., *Process for isolating a protein composition from a muscle source and protein composition. 1999.* U.S. Patent No. 6005,073.

[242] Hultin, H. O., and Kelleher, S. D., *High efficiency alkaline protein extraction. 2000*. U.S. Patent No. 6136,959.

[243] Hultin, H. O., and Kelleher, S. D., *Process for isolating a protein composition from a muscle source and protein composition. 2001.* U.S. Patent No. 6288,216.

[244] Hultin, H. O., and Kelleher, S. D., *Protein composition and process for isolating a protein composition from a muscle source. 2002*. U.S. Patent No. 6451,975.

[245] Undeland, I., et al., Consistency and solubility changes in herring (Clupeaharengus) light muscle homogenates as a function of pH. *Journal of Agricultural and Food Chemistry*, 20003. 51: p. 3992–3998.

[246] Gehring, C. K., et al., Functional and nutritional characteristics of proteins and lipids recovered by isoelectric processing of fish by-products and low-value fish: A review, *Food Chemistry*, 2011. 124: p. 422–431.

[247] Chen, Y. C., et al., Amino acid and mineral composition of protein and other components and their recovery yields from whole Antarctic krill (Euphausia superba) using isoelectric solubilization/ precipitation. *Journal of Food Science*, 2009. 74(2): p. H31–H39.

[248] Taskaya, L., et al., Texture and colour properties of proteins recovered from whole gutted silver carp (Hypophthalmichthys molitrix) using isoelectric solubilisation/precipitation. *Journal of the Science of Food and Agriculture,* 2009. 89(2): p. 349–358.

[249] Taskaya, L., et al., Functional properties of proteins recovered from whole gutted silver carp (Hypophthalmichthys molitrix) by isoelectric solubilization/precipitation. *LWT – Food Science and Technology*, 2009. 42(6): p. 1082–1089.

[250] Mireles DeWitt, C. A., et al., Protein extraction from beef heart using acid solubilization. *Journal of Food Science*, 2002. 67: p. 3335–3341.

[251] Liang, Y., and Hultin, H. O., Functional protein isolates from mechanically deboned turkey by alkaline solubilization with isoelectric precipitation. *Journal of Muscle Foods*, 2003. 14: p. 195–205.

[252] Kelleher, S. D, and Hultin, H. O. Functional chicken muscle protein isolates prepared using low ionic strength, acid solubilization/precipitation. In *53rd Annual reciprocal meat conference: Meat science in the new millenium.* 2000. Savoy, USA: American Meat Science Association.

[253] Gupta, P., et al., Food Processing Residue- Analysis and its Functional Components as Related to Human Health: Recent Developments, *Austin J Nutri Food Sci*. 2015. 3(3): p. 1068+.

[254] Teixeira, A., et al., Natural Bioactive Compounds from Winery By-Products as Health Promoters: A Review, *Int. J. Mol. Sci.*, 2014. 15: p. 15638-15678.

[255] Djilas, S., et al., By-Products of Fruits Processing as a Source of Phytochemicals, *Chemical Industry & Chemical Engineering Quarterly*, 2009. 15(4): p. 191−202.

[256] Viganó, J., and Martinez, J., Trends for the Application of Passion Fruit Industrial By-Products: *A Review on the Chemical Composition and Extraction Techniques of Phytochemicals, Food and Public Health*, 2015, 5(5): p.164-173.

[257] Laufenberg, G. et al., Transformation of vegetable waste into value added products: (A) the upgrading concept; (B) practical implementation. *Bioresour. Technol.*, 2003. 87: p. 167–198.

[258] Baiano, A., Recovery of Biomolecules from Food Wastes — A Review, *Molecules,* 2014. 19: p. 14821-14842.

[259] Schieber, A., et al., A new process of the combined recovery of pectin and phenolic compounds from apple waste. *Innov. Food Sci. Emerg. Technol*. 2003. 4: p. 99–107.

[260] Kew, T. J., and Berry, R. E., Citrus product color enhancement using extracts of peel of different varieties, *J. Food Sci.,* 1970. 35: p. 436+

[261] Ting, S. V., and Hendrickson, R., Natural color enhancers—Orange peel carotenoids for orange juice products, *Food Technol.*, 1969. 23: p. 947+.

[262] Aravantinos-Zafiris, G., et al., Utilisation of orange by-products-Orange peel carotenoids, *J. Sci. Food Agric.*, 1992. 59: p. 77+.

[263] Stoll, T., et al., Process for the recovery of a carotene-rich functional food ingredient from carrot pomace by enzymatic liquefaction, Innovative *Food Science and Emerging Technologies.*, 2003. 4: p. 415+.

[264] Chen, B. H., and Tang, Y. C., Processing and stability of carotenoid powder from carrot pulp waste, *J. Agric. Food. Chem.*, 1998. 46: p. 2312+.

[265] Tang, Y. C., and Chen, B. H., Pigment change of freeze-dried carotenoid powder during storage, *Food Chem.*, 2000. 69: p. 11.

[266] Mazza, G., Anthocyanins in grapes and grape products, *Crit. Rev. Food Sci.*, 1990, 35: p. 341+.

[267] Pazmino-Duran, E.A., et al., Anthocyanins from banana bracts (Musa _ paradisiacal) as potential food colorant, *Food Chem.*, 2001. 73: p. 327+.

[268] Corrales, M.; et al., Tauscher, B. Extraction of anthocyanins from grape by-products assisted by ultrasonics, high hydrostatic pressure or pulsed electric fields: A comparison. *Innov. Food Sci. Emerg. Technol.,* 2008. 9: p. 85–91.

[269] Castillo, J., et al., Antioxidant activity and radioprotective effects against chromosomal damage induced *in vivo* by X-rays of flavan-3-ols (procyanidins) from grape seeds (vitis vinifera): comparative study versus other phenolic organic compounds, *J. Agric. Food. Chem.*, 2000. 48: p. 1738+.

[270] Gamez-Meza, N., et al., Antioxidant activity in soybean oil of extracts from Thomson grape bagasse, *JAOCS,* 1999. 76: p. 1445.

[271] Katalinic, V., Grape catechins-natural antioxidants, *J. Wine Res.*, 1999. p. 10:15.

[272] Louli, V., et al., Recovery of phenolic antioxidants from wine industry by-products, *Bioresour. Technol.*, 2004. 92: p. 201+.

[273] Benavente-Garcia, O., Castillo, J., Marin, F. R., Ortuno, A., and Del Rio, J. A., 1997, Uses and properties of citrus flavonoids, *J. Agric. Food. Chem.* 45:4505.

[274] Miyake, Y., et al., Isolation of C- glucosyflavone from lemon peel and antioxidated activity of flavonoid compounds in lemon fruit, *J. Agric. Food. Chem.*, 1997. 45: p. 4619+.

[275] Ting, S. V., and Newhall, W. F., The occurrence of a natural antioxidant in citrus fruit, *J. Food Sci.,* 1965. 30: p. 57+.

[276] Lu, Y., and Foo, L., Antioxidant and radical scavenging activities of polyphenols from apple pomace *Food Chem.*, 2000. 68: p. 81–85.

[277] Rodrigues de Sotillo, D., et al., Potato peel waste: stability and antioxidant activity of a freeze-dried extract, *J. Food Sci.*, 1994. 59: p. 1031+.

[278] Rodrigues de Sotillo, D., et al., Potato peel extract a nonmutagenic antioxidant with potential antimicrobial activity, *J. Food Sci.*, 1998. 63: p. 907+.

[279] Chu, Y. H., et al., Flavonoid content of several vegetables and their antioxidant activity, *J. Sci. Food Agric.*, 2000. 80: p. 561+

[280] Suh, H. J., et al., Radical scavenging compounds in onion skin, *Food Res. Int.,* 1999. 32: p. 659+.

[281] Dur, P. D., et al., Antioxidant activity of Mung bean hulls, *JAOCS,* 1997. 74: p. 1059+.

[282] Onyenebo, S.N., and Hettriarachchy, N. S., Effect of Navy bean hull extract on the oxidative stability of soy and sunflower oils, *J. Agric. Food. Chem.*, 1991. 39: p. 1701+.

[283] Moure, A., et al., Natural antioxidants from residual sources, *Food Chem.*, 2001. 72: p. 145+.

[284] Oreopoulou, V., *Extraction of natural antioxidants, in: Extraction Optimization in Food Engineering*, Tzia C., and Liadakis, G. eds. 2003. p. 329–346.Marcel Dekker Inc., New York.

[285] Velioglu, Y. S., et al., Antioxidant activity and total phenolics in selected fruits, vegetables, and grain products, *J. Agric. Food Chem.*, 1998. 46: p. 4113+.

[286] Xing, Y., and White, P., Identification and function of antioxidants from oat groats and hulls, *J. Am. Oil Chem. Soc.*, 1997. 74: p. 303+.

[287] Saint-Cricq de Gaulejac, N., Free radical scavenging effects of anthocyanins in red wine, *Food Res. Int.*, 1999. 32: p. 27+.

[288] Lehtinen, P., and Laasko, S., Effect of extraction conditions on the recovery and potency of antioxidants in oat fiber, *J. Agric. Food. Chem.,* 1998. 46: p. 4842+.

[289] Sainvitu, P, et al., Structure, properties and obtention routes of flaxseed lignan secoisolariciresinol: A review. Biotechnol. *Agron. Soc. Environ.*, 2012. 16: p. 115–124.

[290] Kawada, T. et al., *Production of High Concentration Tocopherols and Tocotrienols from Palm Oil By-Products, 1993.* U.S. Patent 5,190,618.

[291] Gapor, M. T. Palm vitamin E–A value-added tocotrienols-rich fraction (TRF) from palm oil. *Palm Oil Dev. Palm Oil Res. Inst. Malays.*, 1995. 22: p. 7–17.

[292] Gapor, M. T. and Rahman, H. A., Squalene in oils and fats. *Palm Oil Dev. Malays. Palm Oil Board,* 2000. 32: p. 36–40.

[293] Hajkowicz, S., *Global megatrends: Seven patterns of change shaping our future, 2015*. CSIRO Publishing.

[294] Augustin, M. A., et al., Role of food processing in food and nutrition security, *Trends in Food Science & Technology*, 2016. 56: p. 115-125.

[295] Hinrichs, C. C., Transitions to sustainability: A change in thinking about food systems change? *Agriculture and Human Values*, 2014. 31: p. 143-155.

[296] Lowe, P., et al., Socio-technical innovation for sustainable food chains: Roles for social science. *Trends in Food Science and Technology,* 2008. 19: p. 226-233.

In: Food: The Vital Ingredient ISBN: 978-1-53614-402-4
Editors: A. B. Hanley and W. R. Russell

Chapter 9

CONSUMER AWARENESS OF SCIENCE AND TECHNOLOGY RELATED TO FOOD RETAIL AND MASS PRODUCTION

A. Bryan Hanley
Knowledge Transfer Network, London, UK

INTRODUCTION

Modern diets in the UK have been shaped by a number of events – both technological and societal. Advances in preservation techniques – particularly canning and freezing – allowed food to be eaten out of season. Globalisation of food production and transportation has accelerated this trend. This has led to an apparent increase in choice of ingredients/food products that can be consumed at any time of the year. In addition the concept of food for health has been underpinned by advances in nutrition and a better understanding of the link between diet and chronic disease.

While the 5-A-DAY message has been heavily promoted in the UK, for both men and women, the proportion consuming five or more portions per day increased from a level of 13% for men and 14% for women in 2001 (when measurements started) reaching a peak in 2006 at around 28% for men and 32% for women. Levels have dropped since then. The most recent National Diet and Nutrition Survey data shows that 27% of adults, 35% of older adults and 8% of 11-18 year-olds achieved the 5-A-DAY recommendation. Total vegetable purchases per household have fallen from 2.9 kg per week in the 1940s to 1.9 kg in the 2000s. In addition, more recently, purchases of vegetables peaked in 2005 but by 2014 had fallen 5.2 per cent since 2007. Lowest income households (decile 1) purchased 6.6 per cent less vegetables in 2014 than in 2007. Reduction in vegetable purchases took place in the 1950s and 60s and plateaued at around 2.4 kg per household per day in the 60s, 70s and 80s. A significant further reduction occurred in the 1990s and 2000s to the level of 1.9 kg in the latter period [1].

One factor which has impacted on dietary choice, and on vegetable consumption, has been the increase in fast food outlets and eating out of home generally. For example, the number of fast food outlets rose by 45% in Norfolk between 1990 and 2008 [2]. This growth in apparent food choice does not seem to have been accompanied by an increase in healthy eating. This may be because the options made available through fast food (and other) outlets are intrinsically less healthy. Most food was eaten and prepared at home in the 1950s and 60s. Typically, meals tended to have the same constituents on the same day of the week. So, for example, in England a roast dinner on Sunday was typical followed by cold meat (left over from Sunday) on Monday. For the rest of the week, meals were prepared following a routine with fish, tinned meats and fresh meat being included in most diets. This lack of variability meant that meals and shopping could be planned and there was no driver for experimentation. In addition, ingredients were limited both in choice and amount.

By the end of the 60s, fast food outlets began to appear in the UK (including hamburger restaurants) and there were increasing numbers of Indian and Chinese restaurants. It was not until the 1970s that MacDonalds made its first appearance on the UK high street. However, most food was still cooked and consumed at home although the increasing availability of packaged and convenience food led to an increase in dietary variation. Instant mashed potato first appeared in 1962 and was followed by others.

There were still comparatively few instant meals and most cooking still required some degree of culinary skill. As the 60s ended and the 70s began, there was an increase in a major convenience aid – refrigeration. By the end of the 70s, two out of every five households had a freezer and most homes had a fridge. This changed both cooking and shopping habits.

Another trend is the increasing use of generic names for types of food. Whereas in the past people would describe what they ate in the context of its ingredients (chicken, lamb, potatoes etc.). This was abbreviated into terms to describe the meal itself (e.g., roast, hot pot, pie). When non-native foods began to be introduced, it was relatively easy to use the same type of nomenclature (e.g., curry, pizza, moussaka). The situation has reverted to describing a style of cooking or its geographic origin (e.g., going for a Chinese/Indian/Greek/Italian). This takes us further away from an understanding of what the constituents of a meal are in terms of ingredients and how they are combined and cooked.

One way of examining home cooking habits is to study the daily energy intake of adults rather than spending patterns. In the US a recent examination of NHANES data and energy consumption suggests that in-home consumption has decreased from >90% in the 60s to <70% in 2008 [3]. The same study found that, in the latest set of figures, despite consuming 72% of daily energy from home food sources, lower income individuals appear to be increasingly less likely to cook, suggesting a greater reliance upon foods that requiring little preparation rather than increased eating out. The inability to cook has been cited as reason for the trend towards less cooking and more assembling of food [4]. In the UK, spend on eating out in 2014 was £12.40 per person per week compared to £29.57 spent on in house consumption and this was 3.9% higher than in 2011 [5].

According to the Organisation for Economic Co-operation and Development (OECD) the US population attaches on average little importance to cooking relative to the other surveyed countries. The US is also one of the countries where relatively little time is spent eating as a primary activity and where obesity rates are amongst the highest in the OECD [6]. This fits a pattern of less cooking, less eating time and low regard for food combining to lower discriminatory capability between different foods and an emphasis on pre-prepared meals rather than an awareness of ingredients.

Given the increase in consumption of convenience foods, there is a need to develop further the healthier options for consumers at an affordable price. In addition, growing populations and increasing wealth in Asia and other markets is leading to an increased demand for animal protein. The WHO have estimated that per capita consumption of animal protein will increase globally by 50% by 2030 and by > 80% in Asia. Others have forecast an annual growth in consumption of >10% in Asia compared to 5% globally. Asia cannot meet such demand from internal resources [7]. By 2050 the global population is expected to rise from 7.4 billion today to 9.6 billion and will require 70 per cent more protein than is currently available.

Food Security, including the provision of animal protein, is a key policy driver for the UK. Food security has many dimensions and responsibility for it rests with at least two departments—Defra, and BEIS. As the report of the House of Commons Environment, Food and Rural Affairs Committee on Food Security noted in 2014, the pressure on animal protein will have a knock on effect not only on animal production itself, but on crops used for animal feed – particularly soya. Furthermore the report concluded "We are concerned about the potential impact of projected rising trends in global demand for animal protein on the price of animal feeds and the cost of production" [8].

Under these pressures, the need to develop further consumer friendly, technologically based solutions becomes ever more of an imperative.

The meat free market in the UK was in excess of £650M in 2014 and continues to rise. However it is not just meat free. Recent surveys by Mintel suggest that up to 20% of consumers are positive about switching from a purely meat protein diet to one containing a mix of meat and vegetable or other protein. In addition, there are significant opportunities to develop novel technologies in areas such as additive manufacturing, crop husbandry, diet and health and industrial biotechnology for food production where the UK already has significant strengths.

Science and Technology

When considering the combination of pressures on food supply and on food security and the increasing burden of chronic, diet-related disease on global societies, the science and technology that will be required must address challenges and be both business and consumer led rather than technology led.

The primary challenges may include:

- An understanding of the more efficient utilisation of protein and of metabolism including factors influencing the metabolism of food.
- An analysis of the factors which govern food choice. This would require a better understanding of the neurology and neuroscience which governs human behaviour and will include the investigation of the neuro-computational basis of decision-making. A better and more complete understanding of oral and gut receptors and their interactions with each other and with the brain. Programming of taste and other key responses.
- An analysis of efficient production methods in farming and food processing including the development of better and more efficient methodology around resource reutilisation, particularly water, energy and key nutrients such as phosphate.
- The underlying soil science, soil mechanics and optimised production systems.
- Fundamental understanding of the soft matter and physics of food structuring and how it can predict the retention and delivery of key nutrients including the triggered release of micro nutrients in the gut.
- Modelling to understand the more efficient use of ingredients/food materials – particularly in the context of novel food sources and processing for production of textured foods with both improved sensory characteristics and high consumer acceptability.

However each new technology must also be accepted both by industry in terms of harvesting, yield and processability and by the consumer in terms of acceptability and cost. In the case of the latter, there are sometimes significant socioeconomic hurdles that have to be overcome and, even when it appears that they have been, things can change.

Perceptions about food are fluid and influenced by a number of factors. This is perhaps illustrated by the story of the genetically modified tomato (Flavr Savr). The tomato had a deactivated gene which was turned off using anti-sense technology. This gene coded for polygalacturonase, an enzyme involved in fruit ripening, and this meant that tomatoes could be ripened on the vine and would still have a long shelf life. A variation of this technology was developed by Zeneca and used to produce tomato puree, which was on supermarket shelves in the UK between 1996 and 1999. The fact that it was 20% cheaper than conventional puree helped to drive sales. After significant lobbying by a number of groups, public perceptions changed and the supermarkets which stocked the genetically modified (GM) puree decided to remove it from their shelves. The House of Commons Science and Technology Committee noted this change in public perception regarding GM foods [9]. Public perception was greatly influenced by scare stories, which talked in vivid detail about 'Frankenstein food' and other potential problems – many of which were largely unsubstantiated. Beliefs about the safety, convenience, healthiness, flavour, satiating capacity and cost of food can all influence food choice – even if these beliefs are not necessarily always rational.

There is a need to develop new foods that fit with the current convenience food, simple choice paradigm (quick to prepare, simple to cook and serve) but which are healthy, nutritious and contain components that are available in multiple formats including the meal choices that are commonly made by consumers. This will require new technologies (e.g., additive manufacturing, novel storage and cooking methods) underpinned by a better understanding of diet and health and of how consumers think and approach decisions about what they choose to eat. One ongoing challenge will be to provide foods for people in a format they recognise and which contain increasing amounts of healthy ingredients (particularly vegetables) but taste the same as the convenience foods they are used to. This, together with the pressures on the food supply mentioned above, will create an environment of accelerated progress and some new and distinct food systems based on technologies that are still in the process of being developed.

Despite the challenges associated with communication of technology to consumers illustrated in the case of GM, the local and global challenges associated with providing a safe and secure food supply will be strong drivers for the development of new technologies.

Consumer Acceptance of Technology

Key to all of this is consumer communication and acceptance of new technologies. This will be illustrated in four areas.

1) Genetic and related technologies
2) Increased yields and crop protection
3) Novel processing technology
4) Reduction of waste and greater sustainability.

Genetic and Related Technologies

Consumer resistance to genetic and related technologies remains high. This was in part caused by the strong campaigning by groups such as Greenpeace. There has, however, been something of a response to the Greenpeace stand with 107 Nobel Laureates recently protesting against the Greenpeace position. In particular, the scientists called Greenpeace out for publicly opposing golden rice - a GM crop that it has been claimed has the potential to save millions of lives each year by reducing vitamin A deficiency in the developing world. The letter from the Nobel Laureates comments on the Greenpeace position. "They have misrepresented [GMOs] risks, benefits, and impacts, and supported the criminal destruction of approved field trials and research projects," the researchers write in their letter. “How many poor people in the world must die before we consider this a ‘crime against humanity’?”

While a number of variations on the theme of introducing genes into crops have become main stream, new technologies have also been investigated. One of these is the CRISPR-Cas9 system. Essentially this is a gene editing technique and introduces a new gene or genes using a DNA cutting enzyme called Cas-9 coupled to a specific sequence finder – a piece of RNA called guide RNA (gRNA). The system has been used successfully to edit the genome of rice and citrus fruit. Greenpeace have already expressed their opposition to gene editing and regard species modified using such techniques to be just another form of GM however plant scientists say that new editing tools such as CRISPR–Cas9, involve no more than making tiny, precisely targeted changes to a gene that are indistinguishable from natural mutations. The EU is not expected to make any decision as to whether such modified plants should be subject to GM regulations until 2018. However the In a statement in July, the Court of Justice (ECJ) said it "takes the view, first of all, that organisms obtained by mutagenesis are GMOs [genetically modified organisms] within the meaning of the GMO Directive". The potential consequences of this opinion are significant for technology in Europe and elsewhere.

In the meantime, work continues in China and elsewhere to use gene editing to modify existing crops and its application in animals and for human use and some countries such as Sweden and Finland have decided that plants produced using gene editing technology should not be restricted by application of the GM regulations.

A lot will depend upon the acceptance or otherwise of the arguments of scientists active in the field and the counter-arguments being made by Greenpeace and similar groups.

Increased Yields and Crop Protection

The drive for increased productivity is in response to a clear need for increased raw material production. This includes increasing yields from current crops and animal husbandry practices, growing plants in animals in places where they are not currently able to survive and thrive and the use of new sources of protein, carbohydrate and fat (e.g., algae, conversion of current non crop plants, fungi etc.). In part this can be facilitated by the use of GM or gene editing techniques however, as we have seen, the regulatory position remains unclear – at least in Europe. While many consumers are in favour of free range and organic produce, the price premium such products attract make them less popular than the conventionally grown alternatives. In addition, intensive farming requires the use of various crop protection production. Organisation such as the Soil Association have pointed towards a growth of 7.5% in the value of sales of organic produce in the UK in 2016 as a sign of the growing popularity of foods produced in this way. However it should be noted that sales are still lower than they were at their peak in 2008 – before the recession. It should be noted, however, that organic accounts for just 1.4% of the UK food and drink sector and the amount of land under organic cultivation has fallen for three years in succession. In addition, non-organic sales have diminished in part because of the growth of discount supermarkets and severe discounting of products.

Yields of food crops and animals for consumption have increased in countries with highly developed farming systems. For example, according to figures from DEFRA, the index for total sector productivity in the UK increased from 100 in 1972 to 152 by 2014. According to the FAO In the past 50 years, global crop production has expanded threefold. This increase has been driven largely by higher yields per unit of land, and crop intensification resulting from multiple cropping and/or shortening of fallow periods. The expansion of arable land area allocated to crops has played a less important part in production increases. However the same organisation estimates indicate that to meet the projected demand, global agricultural production will have to increase by 60 percent from its 2005–2007 levels.

A major source of loss of crops is pest damage however there are increasingly strict regulations regarding the use of crop protection agents and consumers have a degree of suspicion about 'chemicals' being sprayed onto plants (with, it has been suggested, deleterious effects on beneficial insect populations such as bees) and the prophylactic use of antimicrobials in animals. The costs of pesticide treatments of some major crops are given below.

Wheat

In Australia the average total cost of treatment for invertebrate pests in wheat nationally is $77.9 million or $6 (€4) per ha grown. If we assume a similar figure globally then the cost would be around $1.3Bn which equates to €900M in total for global wheat treatment. Given that the value of the crop is around €80 bn then the cost of invertebrate crop protection is around 1.5% of the value of the crop as a whole.

Rice

Figures for rice are more difficult to judge since the cost of insecticide treatment varies globally however in the US insecticide treatment costs $9 per hectare. This suggests a global cost of $1.3bn which represents 0.7% of the value of the crop.

Corn

Figures from the US (Illinois) suggest that insecticide use for corn costs $24 (€21) per hectare. The total global cost would then be €3bn. With a total crop value of €50bn, the cost of insecticide treatment is about 6% of the total value of the crop.

Cassava

In the case of cassava, it has been estimated that pesticide treatments cost around €20 per hectare. This gives a total global cost of €400M. With a total value of €24bn, the cost of insecticide treatment is about 1.6% of the value of the crop.

Given these figures and the difficulty in growing crops without some sort of pest management system, there is an urgent need to introduce new methods of control of pests. Some of the most interesting areas are around the use of peptide-based agents which target the neuroendocrine system of pests specifically with no damage to other insects and which break down rapidly in the environment to harmless products. Further work will be needed and it is important that the techniques used (and their advantages) can be explained to consumers, regulators and, not least, politicians.

Novel Processing Technology

The history of food processing is filled with examples of new and better processing technology. While there have been significant advances – many of which have helped to make food safer, more palatable and more interesting, there have also been some significant failures. Food irradiation was thought to be a considerable step forward in food safety and in maintaining the quality of food. However it became linked in the consumers' minds with radioactivity and so was never adopted by the food industry on the scale that might have made a difference, and been profitable.

One of the most recent technologies to come from adjacent technology areas is additive manufacturing. Additive manufacturing (or 3D printing as it is commonly referred to) is a process of creating objects by making successive layers. It has commonly been used in process industries to allow manufacturers to carry out fast prototyping and to thereby create pieces of equipment at lower cost and with considerably reduced timescales.

In principle additive manufacturing could be 'feedstock blind'. What this means is that, from any combination of macro and appropriate micronutrients and constituents, an additive manufacturing facility could create a number of desired foods. So the substitution of plant for animal protein or of rice for wheat, for example, would not require the use of a whole different processing facility. This means that a range of raw materials could be used to create foods. This could clearly have significant implications for processing. A second and related aspect is that such facilities may not require the complex infrastructure that current food processors need. Using a 3D printer as a model, it is feasible that food could be produced in technologically and infrastructure-poor areas and this would encourage in time and at place processing which in turn would affect traditional processing paradigms, transport, refrigeration etc.

This leads directly onto the second opportunity. The idea of creating foods as and when required by using additive manufacturing has the potential to revolutionise current food processing, storage and distribution. When this is added to the ability to create foods with specific properties (e.g., with a targeted range of micronutrients and other components based on age, nutrigenomic profile, existing or future conditions and vulnerabilities etc.) then foods become personalized and specific. Whether this personalization occurs at the supermarket, at a restaurant or even in the home will be dependent upon the degree of complexity of the food being created and the availability of the various raw materials.

While the additive manufacturing process has great potential, there are a number of challenges. Current systems for additive manufacturing have, on the whole, been developed to make use of inorganic materials in the manufacturing process. The main focus for additive manufacturing research in the UK, for example, is in metals technology and polymers. In life sciences, orthodontic research is a major target. The use of additive manufacturing to make complex materials with labile starting materials such as foods is rare. In addition, many additively manufactured products are used to create equipment rather than ingested products. This tends to mean that they are homogenous. The technology for the production of multi-heterogeneous foods which would have to be produced by a process which included layering, processing and layering again, is not currently available.

Solving these technological problems while at the same time explaining the technology to the consumer will be essential if additive manufacturing is to fulfil its potential in the production of food.

Reduction of Waste and Greater Sustainability

The production of waste and the imperfect utilisation of both raw materials and finished products is a major challenge for the agri food industries. The eighth report of the House of Commons Environment, Food and Rural Affairs Committee on food waste [10] noted that "Food waste is a global public policy issue. The Food and Agriculture Organisation of the United Nations (FAO) estimates that one-third of food produced for human consumption is lost or wasted globally, which amounts to approximately 1.3 billion tonnes per year. In the UK, it is estimated that 10 million tonnes of food and drink waste arises post-farmgate each year, 60% of which could be avoided. This has serious economic, environmental and social implications and impacts.

The Committee made a number of recommendations to try to address the problem. Perhaps the most consumer relevant of these was the consideration to remove 'best before' as a label for food and to simplify the system by having a 'use by' designation only. This would remove a major source of confusion for consumers who assume (wrongly) that a best before designation means that food cannot safely be consumed after that date.

The use of anaerobic digestion as a default final stage for waste food has been widely criticised and alternative ways of making use of both surplus food and food processing waste products has broad support. Apart from any ecological considerations, it is also more economically advantageous to transform food and its components into more complex, higher value endpoints (e.g., flavours, colours etc.) rather than to reduce complex molecules to the simplest hydrocarbon – methane. That loss of complexity and potential value will require significant chemical effort over the coming years. The potential to explain to consumers that waste is actually potentially of great value is significant.

CONCLUSION

Consumers have a close and intimate relationship with the food that they eat. The ingredients of food are of prime importance to how that food looks and tastes. Food is safer now that it has ever been however its very attractiveness combined with lower cost and greater availability and choice has contributed to over consumption. When a loss of physical exercise and a more sedentary work and leisure lifestyle is added in, the result is the current situation where over half the adult population in many developed countries are overweight or obese. Chronic disease as a result of this is a major challenge for individuals, populations and society.

If this were to be considered in isolation then the solution could include using chemistry to underpin the production of healthier foods with ingredients more attuned to a healthy lifestyle and processed in ways that retain the attractiveness for consumers but enhance the beneficial properties of all of the ingredients.

However demographics and changing population and consumption dynamics suggest that overlaid on this will be a need to create more of one of the basic building blocks of a healthy life – protein. Animal protein is likely to face mounting pressures as increasing numbers of people have to be fed – not simply to avoid starvation as in the past, but because they aspire to the higher animal protein diet they see being consumed in North America and Europe.

Part of the solution will be changes and improvements in primary production and shifts in diet to a lower animal-based consumption pattern. Greater emphasis on the elimination of waste will also provide some breathing space as will protein from new (and some older) sources (insects, laboratory grown meat substitutes, algae, mycoprotein). However all of this will present challenges in understanding the chemistry of food and its components, its physical make up and new and better processing techniques underpinned by science.

However ultimately it is the consumer who will be the final judge of new technologies and the food produced by them. This is one reason why insect protein will most likely remain a small part of the answer and laboratory grown meat will be treated with some degree of suspicion. It is incumbent upon chemists, in particular, to ensure that the advances they are able to engineer are understood by consumers and that technology does not go down the route of food irradiation or GM.

References

[1] MAFF (Ministry of Agriculture, Fisheries and Food) (1991) Fifty Years of the National Food Survey 1940–1990. HMSO: London, *Family Food* 2014, DEFRA, 2015.

[2] *Area deprivation and the food environment over time: A repeated cross-sectional study on takeaway outlet density and supermarket presence in Norfolk, UK, 1990–2008.* Eva R. Maguire, Thomas Burgoine, Pablo Monsivais Health Place. 2015 May; 33:142-7.

[3] Smith, LP, Shu Wen Ng, and Popkin, BM, Trends in US home food preparation and consumption: analysis of national nutrition surveys and time use studies from 1965–1966 to 2007–2008. *Nutr J.* 2013; 12: 45.

[4] Winkler E, Turrell G. Confidence to cook vegetables and the buying habits of Australian households. *J Am Diet Assoc.* 2010; 110: S52–S61.

[5] *Family Food 2014,* DEFRA, 2015 https://www.gov.uk/government/collections/family-food-statistics.

[6] OECD (2011), *Society at a Glance 2011 - OECD Social Indicators* (www.oecd.org/social/societyataglance2011.htm).

[7] Wu et al. *Ann NY Acad Sci.* 2014 Aug;1321: 1-19.

[8] Environment, Food and Rural Affairs Committee, Second Report of Session 2014-15, *Food Security,* HC 243.

[9] House of Commons Science; Technology Committee (May 18, 1999). *Scientific Advisory System: Genetically Modified Foods.* Retrieved May 4, 2011.

[10] House of Commons Environment, Food and Rural Affairs Committee, Eighth Report of Session 2016-17, *Food Waste in England,* HC 429.

In: Food: The Vital Ingredient ISBN: 978-1-53614-402-4
Editors: A. B. Hanley and W. R. Russell

Chapter 10

FOOD WASTE AND BY-PRODUCTS VALORIZATION

Dimitris Charalampopoulos
University of Reading, Reading, Berkshire, UK

FOOD WASTE IN THE SUPPLY CHAIN: SOURCES AND IMPACT

The world is facing significant food and nutrition security challenges in the next decades due to the scarcity of natural resources, climate change, and the increasing global population which is expected to reach 9 billion by 2050 [1]. In the European Commission High Level Conference entitled "FOOD 2030: Research and Innovation for tomorrows' nutrition and food systems" that took place in October 2016 it was highlighted that a food systems approach needs to be adopted that is underpinned by sustainability and encompasses the entire food value chain. The consensus is that in order to address these challenges our food systems need to be future-proof, i.e., more sustainable, resilient, responsible, diverse, competitive and inclusive [2]. To succeed in this, a number of approaches could be taken, including increasing total food production, improving access to food and distribution channels, improving the nutritional value of food and the diets of consumers, reducing food waste and stimulating the circular economy through the valorisation of unavoidable food waste and low value by-/co-products. According to the Food and Agriculture Organization of the United Nations (FAO) approximately 1/3 of the food produced globally on an annual basis is wasted across the food supply chain, amounting to approximately 1.3 billion tons [3]. However, the available data on food waste have many gaps due to the inconsistent definitions of food waste and the methodologies used for calculation. A study published by the European Commission in 2010 [4] showed that the food processing waste generated by the 27 EU member countries amounts to 90 million tons per annum, which was broken down by sector to households (42%), manufacturing (39%), food service/catering (14%) and retail/wholesale (5%). The most comprehensive national data on food processing waste are provided by the Waste and Resources Action Programme (WRAP) in the UK. It was shown from a 2010 study that about 2.5 million tons of food processing waste are generated each year in the UK compared to about 8.3 million tons of household waste; in addition, about 1.5M tons of low value by-products (e.g., distillers' grains, confectionary by-products, sugar beet pulp) go into animal feed [5] Food waste includes materials which are lost during primary production and secondary processing, packaging, losses during storage and distribution, as well as those discarded by consumers. Food processing waste and by-products are mainly derived from plant and animal sources; plant wastes include vegetable and fruit trimmings (comprising whole/partial organs) and cereal residues such as bran and spent grains though to pulps and residues remaining after oil, starch, sugar and juice extraction. Animal wastes include blood, bones and tissues of meats and fish, and solid and liquid wastes from the dairy industry. More complex mixed wastes are produced in the production of ready meals which are composed of both animal and plant material. A review by Raak et al. [6]. looks into detail into the processing and product related causes for the generation of waste across the food supply chain including overproduction, processing defects and logistical causes, whereas a systematic review by De Steur et al. [7]. compiled a number of studies applying value street mapping and other lean tools, such as Just in Time and 5S, aiming to identify approaches that can be used to reduce waste in food supply chains. Another review by Mercier et al. [8] assessed the state of the commercial cold food chains and identified weaknesses and development opportunities to minimise waste generation.

Given the large amounts of food waste generated annually, its organic nature and high biological oxygen demand (BOD) and chemical oxygen demand (COD), food waste has a significant environmental, societal and economic impact. In terms of greenhouse gas (GHG) emissions, the overall impact of food waste in the 27 EU countries was estimated in 2010 as 170 Mt of CO_2 equivalent per year; this was projected to increase to 240 Mt of CO_2 equivalent per year by 2020.[4] Moreover, a recent study estimated that emissions associated with food waste are likely to increase substantially to around 1.9 and 2.5 GT annually by 2050, as a result of the increased growth population [9] and most likely the expansion of the middles classes in developing countries. According to this study up to 14% of emissions could be avoided by managing food use and distribution better. In terms of potential economic benefits, reduction in food waste will have significant benefits as the direct economic loss of food waste has been estimated to be around $750 billion annually.[3]

WASTE RECOVERY AND VALORISATION STRATEGIES

As food waste is a significant issue in the EU, the EU Waste Framework Directive was adopted by the EU member states in 2008 (2008/98/EC), which introduced a hierarchy of food waste management strategies (Figure 1); this prioritises the prevention of food waste generation and is followed by preparing for re-use and recycling of waste materials (e.g., energy, chemicals), other recovery methods (e.g., composting, animal feed), and finally disposal to landfill as the least preferred method [10]. As part of the waste prevention strategy, a wide range of initiatives have been investigated at all levels (national, international) by various stakeholders (e.g., public authorities, businesses, non-governmental organisations, consumer groups) including awareness campaigns, such as the Love Food Hate Waste campaign in the UK and the Think.Eat.Save campaign which is supported by the United Nations Environment Programme (UNEP) and FAO, information tools (guides, brochures), training programs, logistical improvements, regulatory measures, and food redistribution programmes [11]. In terms of recycling, re-use and recovery, a number of approaches are applied at commercial scale, including the utilisation of food waste and processing by-products as animal feed, the composting of food waste and the production of biogas through anaerobic digestion (AD) [12-14]. AD has been a popular waste recovery strategy, particularly in Germany; moreover in the last few years this strategy has been significantly promoted in the UK through government subsidies. The most recent figures from the UK Anaerobic Digestion and Bioresources Association (ADBA) quote 540 AD plants operating in the UK in 2017, generating over 700MW of power. The demand in AD is still strong although it is expected to decrease as the UK capped the amount of subsidies in 2016 [15]. Considering the above strategies and the advances in the last 20 years in the area of biofuels and industrial biotechnology, there is significant potential for the development of novel value chains from food waste materials, targeting the production of added value products, such as biofuels, platform and specialty chemicals, and biomaterials; these can find applications in a range of industries including the chemical, food, agriculture, and cosmetics and personal care industries [16-18]. In order for such a strategy to be commercially viable though, it needs to be integrated within a food waste biorefinery, where low value (e.g., animal feed, compost) and higher value products are produced. This will ensure that most organic components of food waste are fully recovered and that there is a significant environmental impact and an economic impact through the stimulation of the circular bioeconomy [19, 20]. The bioeconomy is a term that was coined by the European Commission in 2012 to exploit the possibilities of converting renewable biological resources from land and sea, as well as waste, as inputs to food and feed, industrial and energy production [21]. The strategy has three pillars including: (a) Investment in research, innovation and skills for the bioeconomy, (b) Development of markets and competitiveness in bioeconomy sectors and (c) Reinforced policy coordination and stakeholder engagement, through the creation of a Bioeconomy Panel, a Bioeconomy Observatory and regular Stakeholder Conferences. To address the R&D challenges in these areas, the Commission developed the strategic challenge 'Food security, sustainable agriculture, marine and maritime research, and the bioeconomy' under the programme Horizon 2020, providing funding in excess of €5 billion to academia and industry. The significant potential of the bioeconomy has also been highlighted by the UK Science and Technology Select Committee [22] and the Unites States Department of Agriculture (USDA) [23]. To succeed in implementing such a strategy though, it is essential that we build a dynamic academia/industry research base, enhance our knowledge on food waste composition, explore product applications with high market potential and define the functional requirements of such products. In this respect, it is important to understand how to apply existing or emerging technologies for extracting, fractionating and converting components from food waste feedstocks to products, at large scale and in an efficient, environmentally friendly and cost-effective way.

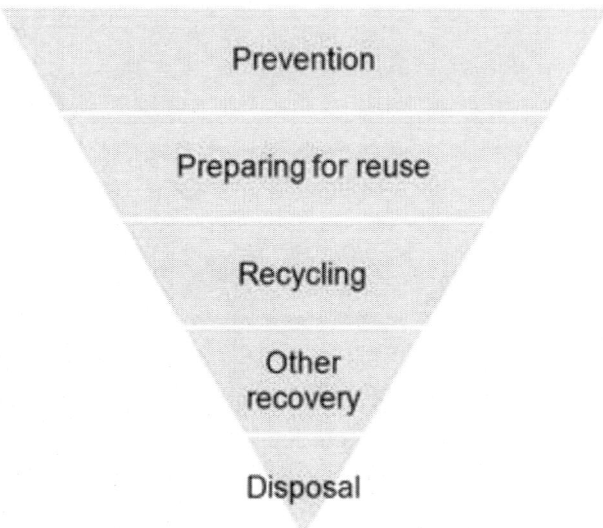

Figure 1. Hierarchy of food waste minimisation strategies.

Case Study: Valorisation Strategies for Dried Distillers' with Solubles (DDGS)

A case study for the valorisation of a food processing by-product is DDGS, a by-product from the bioethanol and distillery industries that is produced in large amounts worldwide. With regards to bioethanol production, approximately one third of every kilogram of wheat is converted into ethanol, one third into carbon dioxide and one third into DDGS [24]. It can be derived from various starchy-rich grains such as corn, wheat, sorghum and barley, a variation that may result in dissimilarities in the chemical composition and physical properties of DDGS. In the USA, DDGS from bioethanol plants is corn-based due to the predominant usage of corn as feedstock for fuel ethanol production, whereas in Canada, the UK and most of European countries, DDGS is mainly wheat based [25, 26]. DDGS is used as animal feed and is an important source of energy, protein, water-soluble vitamins and minerals for livestock. The addition of DDGS to livestock feed can account for up to 30% of the diet on a dry matter basis, as higher levels may cause palatability and excessive protein consumption issues [27]. Moreover, the compositional variation in DDGS (source and process dependent) is an obstacle for its use as an animal feed supplement for ruminants [28, 29]. Considering the above and the increased bioethanol-derived DDGS availability that is expected in the future, there is an impetus to identify alternative ways to enhance the value of DDGS and further support the circular bioeconomy concept.

DDGS consists to a large extent of organic macromolecules including soluble and insoluble dietary fibre, (ca 40-50% w/w, mainly arabinoxylan, cellulose, and β-glucans), proteins (ca 25-40% w/w) and oil (ca 3-12%) [24]. Due to the absence of a rigid lignocellulosic structure, DDGS is amenable to relatively mild processing, which can lead to the production of several value-added streams; these can act either as end-products or starting materials for secondary processing. A schematic representation of a conceptual valorisation strategy based on DDGS leading to multiple products (from low to mediu/high value) is depicted in Figure 2; these include biofuels, biopolymers, platform chemicals, prebiotic oligosaccharides [30, 31] and packaging materials. These products could be derived by effectively exploiting the two main DDGS components, i.e., carbohydrates and proteins, accounting approximately 70% of the total composition. From a technological perspective, it is possible to achieve the separation and recovery of proteins and carbohydrates using established methodologies that have good scalability potential (alkali or solvent extraction) with good recovery yields ranging from 40-80% [32-35]. From an economic perspective, given the fact that DDGS has currently a market value as animal feed (the price of which normally fluctuates as a result of the wheat and corn price but normally ranges from $100-$250 per ton), the choice of product(s) deriving from DDGS should have a higher value than the feed value, as well as strong market volume and potential in order to compensate for the additional processing, materials and energy costs. Research thus far has demonstrated that such a biorefinery approach for DDGS is feasible at the laboratory and in some cases pilot scale although more translational research coupled with detailed process economics is needed to achieve commercial realisation and exploitation.[24]

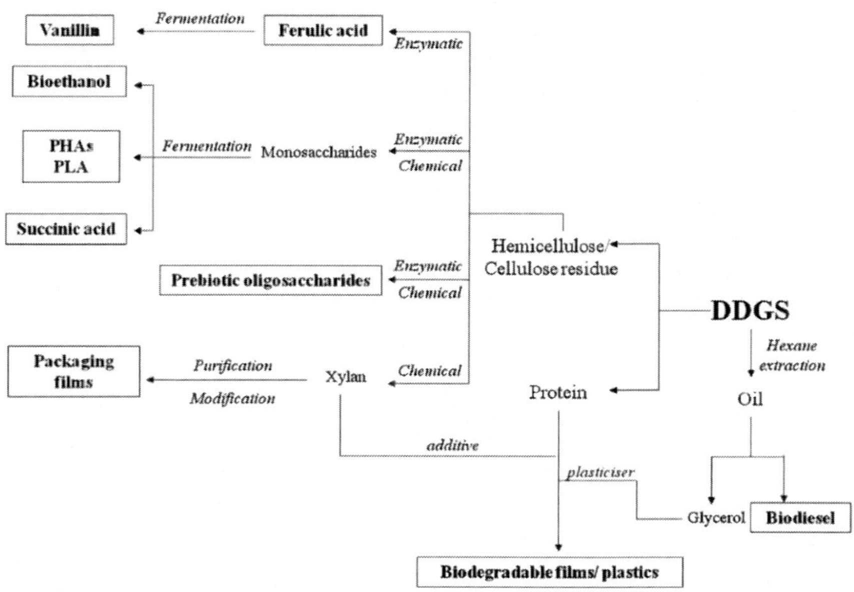

Figure 2. Conceptual valorisation scheme for DDGS.

R&D Challenges for Food Waste and By-Product Valorisation

Despite the fact that food waste and by-products are available in high amounts, are rich in exploitable components (e.g., protein, carbohydrate, phytochemicals, minerals, etc), and are generally simpler to handle than lignocellulosic material due to the absence of lignin, there have not been any significant commercial successes implementing a food waste biorefinery approach. The main reasons are that this approach is considered by businesses a high-risk strategy mainly due to the large investment required, unpredictable return of investment, regulatory hurdles and technical challenges (e.g., heterogeneity of food waste feedstocks, logistical issues, complex manufacturing processes). However, there has been a momentum over the last few years, driven by environmental and consumer pressures, legislation, and the market potential of such bioproducts, for the development of industrial uses of functional components derived from food waste and by-products. It is clear though that food waste valorisation is a challenge that needs to be addressed through an inter-disciplinary approach involving a range of disciplines (e.g., chemistry, microbial fermentation, enzymology, process engineering and design, process economics, product formulation, marketing, regulation, etc) and a range of stakeholders including a diverse academic research base, primary producers, food manufacturers, chemical and industrial biotechnology manufacturing specialists, and end-users (food and feed suppliers, chemical, cosmetic and personal care companies), in order to achieve significant economic, social and environmental impact. There have been a number of activities over the last few years across the EU supporting the concept; examples include: (i) Funding from the Bio-based Industries Joint Undertaking (BBI-JU), a public-private partnership between the European Commission and the Bio-based Industries Consortium (https://www.bbi-europe.eu), (ii) an EU COST action (large EU Network) on 'Food Waste Valorisation for Sustainable Chemicals, Materials and Fuels (EUBis)' (http://www.costeubis.org), (iii) FoodWasteNet (http://www.foodwastenet.org), an academic-industry

Network funded by the UK Biotechnology and Biological Sciences Research Council (BBSRC) and (iv) the Courtauld Commitment 2025 led by WRAP UK, a voluntary agreement that brings together organisations across the food system – from producer to consumer – to make food and drink production and consumption more sustainable (http://www.wrap.org.uk/content/courtauld-commitment-2025).

There is therefore a need to expand collaborative activities in this area to develop novel approaches based on innovative biological and process engineering that lead to commercial processes and novel or improved products. A number of challenging research areas are listed below where more R&D activities are needed:

Identify food waste and low value by-product streams at a national/regional level offering the highest potential for valorisation taking into account their composition, variability, volume, seasonality and storage stability.

Improve through primary processing the stability of food waste and processing by-products.

- Explore innovative processing for the separation and functionalisation of carbohydrates/proteins from food waste and by-products, which have potential in food, packaging or personal care products as thickeners, gelling agents, stabilisers, and biodegradable packaging materials.
- Increase the activity of enzymes used for fractionation or hydrolysis, which are not as effective against complex raw materials such as food waste.
- Improve on the low conversion yields of pre-treated food waste to products by microbial fermentation (for the production of platform or specialty chemicals and biofuels).
- Increase process yields and productivities by using improved biocatalysts (e.g., engineered enzymes and microbes), process integration (e.g., simultaneous fractionation, hydrolysis and biotransformation of food waste) and novel engineering designs for upstream and downstream processing (e.g., continuous systems).

- Develop process designs for novel value chains based on food waste and processing by-product streams taking into account the 'roadblocks' (technical, economic, logistical) that businesses are facing in their attempts to put these processes in place and develop joined-up biorefinery solutions.

REFERENCES

[1] United Nations, (2017). *World Population Prospects, Key findings and advanced tables.*

[2] European Commission, 2016. *European Research & Innovation for Food & Nutrition Security, Food 2030 High-level Conference background document.*

[3] Food and Agriculture Organization of the United Nations (FAO), (2011). *Global food losses and food waste.*

[4] European Commission, (2010). *Preparatory Study on Food Waste in the EU27.*

[5] WRAP, 2010. *Waste arising in the supply of food and drink to UK households.*

[6] Raak, K., Symmank, C., Zahn, S., Aschemann-Witzel, J., Rohm, H., (2017). Processing and Product related causes of food waste and implications for the food supply chain. *Waste Management*, 61:461-472.

[7] De Steur, H., Wesana, J., Dora, M. K., Pearce, D., Gellynck, X., (2016). Applying value street mapping to reduce food losses and wastes in supply chains: A systematic review. *Waste Management*, 58:359-368.

[8] Mercier S., Villeneuve, S., Mondor, M., Uysal, I., (2017). Time-temperature management along the food cold chain: A review of recent developments. *Comprehensive reviews in Food Science and Food Safety*, 16:647-667.

[9] Hic, C., Pradhan, P., Rybski, D., Kropp, J. P., (2016). Food surplus and its climate burden. *Environmental Science and Technology*, 50:4269-4277.

[10] *Directive 2008/98/EC on waste (Waste Framework Directive).*

[11] UNEP, (2014). *Prevention and reduction of food and drink waste in businesses and households - Guidance for governments, local authorities, businesses and other organisations, Version 1.0.*

[12] Zhang, C., Su, H., Baeyens, J., Tan, T., (2014). Reviewing the anaerobic digestion of food waste for biogas production. *Renewable and Sustainable Energy Reviews*, 38:383-392.

[13] Li, Z., Lu, H., Ren, L., He, L., (2013). Experimental and modeling approaches for food waste composting: A review. *Chemosphere*, 93:1247-1257.

[14] Martin, D. S., Ramos, S., Zufia, J., (2016). Valorisation of food waste to produce new raw materials for animal feed. *Food Chemistry*, 198:68-74.

[15] ADBA website, http://adbioresources.org/about-ad/ad-map

[16] Matharu, A. S., de Melo, E. M., Houghton, J. A., (2016). Opportunity for high valued-added chemicals from food supply chain wastes. *Bioresource Technology*, 215:123-130.

[17] Ravindran, R., Jaiswal, A. M. (2016). Exploitation of food industry waste for high value products. *Trends in Biotechnology*, 34:58-69.

[18] Turon, X., Venus, J., Arshadi, M., Koutinas M., Lin C. S. K., Koutinas, A., (2014). *Food Waste and byproducts valorization through bio-processing: Opportunities and challenges*, 9:5774-5777.

[19] Dahiya, S., Kumar, A. N., Sravan, J. S., Chatterjee, S., Sarkar, O., Mohan, S. V., (2017). *Food waste biorefinery: Sustainable strategy for circular bioeconomy*, https://doi.org/10.1016/j.biortech.2017.07.176 (in press).

[20] Kiran, E. U., Trzcinski, A. P., Liu, Y., (2015). Platform chemical production from food wastes using a biorefinery concept, *Journal of Chemical Technology and Biotechnology*, 90:1364-1379.

[21] European Commission, (2012). Communication from the commission to the European parliament, the council, the European economic and

social committee and the committee of the regions. "*Innovating for sustainable growth: a bioeconomy for Europe*".

[22] House of Lords, Science and Technology Select Committee, (2014). *Waste or resource? Stimulating a Bioeconomy.*

[23] USDA, (2014). *Why Biobased? Opportunities in the Emerging Bioeconomy.*

[24] Chatzifragkou, A., Kosik, O., Prabhakumari, P. C., Lovegrove, A., Frazier, R. A., Shewry, P. R., Charalampopoulos, D. (2016). Biorefinery strategies for upgrading Distillers' Dried Grains with Solubles (DDGS). *Process Biochemistry*, 50:2194-2207.

[25] Burton, E. J., Scholey, D. V., Williams, P. E. V. (2014). Types, Properties and Processing of Bio-Based Animal Feed. In K. W. Waldron (Ed.), *Advances in Biorefineries: Biomass and Waste Supply Chain Exploitation* (First Edition ed., pp. 771–802). Cambridge, UK: Woodhead Publishing.

[26] Moreau, R. A., Nghiem, N. P., Rosentrater, K. A., Johnston, D. B., & Hicks, K. B. (2012). Ethanol Production from Starch-Rich Cropss Other Than Corn and the Composition and Value of Resulting DDGS. In K. Liu & K. A. Rosentrater (Eds.), *Distillers Grain: Production, Properties, and Utilization* (pp. 103 - 117). Boca Rato, FL: CRC Press: Taylor & Francis Group.

[27] Schingoethe, D. J., Kalscheur, K. F., Hippen, A. R., Garcia. A. D. (2009). Invited review: the use of distillers products in dairy cattle diets. *Journal of Dairy Science*, 92:5802-5813.

[28] Belyea, R. L., Rausch, K. D., Clevenger, T. E., Singh, V., Johnston, D. B., Tumbleson, M. E. (2010). Sources of variation in composition of DDGS. *Animal Feed Science and Technology*, 159:122-130.

[29] Tedeschi, L. O., Kononoff, P. J., Karges, K., Gibson, M. L. (2009). Effects of chemical composition variation on the dynamics of ruminal fermentation and biological value of corn milling. *Journal of Dairy Science*, 92:401-413.

[30] Hutkins, R. W., Krumbeck, J. A., Bindels, L. B., Cani, P. D., Fahey, G., Goh, Y. J., Hamaker, B., Martens, E. C., Mills, D. A., Rastall, R.

A., Vaughan, E., Sanders, M. E. (2016). Prebiotics: why definitions matter. *Current Opinion in Biotechnology*, 37:1-7.

[31] Gibson G. R., Roberfroid, M. B. (1995). Dietary modulation of the human colonic microbiota: introducing the concept of prebiotics. *Journal of Nutrition*, 125:1401–1412.

[32] Xu, W., Reddy, N., Yang, Y., (2007). An acidic method of zein extraction from DDGS. *Journal of Agricultural and Food Chemistry*, 55:6279-6284.

[33] Anderson, T. J., Ilankovan, P., Lamsai, B. P. (2012). Two fraction extraction of α-zein from DDGS and its characterization. *Industrial Crops and Products*, 37:466-472.

[34] Wang, Y., Tilley, M., Bean, S., Sun, X. S., Wang, D.(2009). Comparison of methods for extraction of kafirin proteins from sorghum distillers dried grains with solubles. *Journal of Agricultural and Food Chemistry*, 57:8366-8372.

[35] Chatzifragkou, A., Prabhakumari, P. C., Kosik, O., Lovegrove, A., Shewry, P. R., Charalampopoulos, D. (2016). Extractability and characteristics of proteins deriving from wheat DDGS. *Food Chemistry*, 198:12-19.

In: Food: The Vital Ingredient ISBN: 978-1-53614-402-4
Editors: A. B. Hanley and W. R. Russell

Chapter 11

NEW TECHNOLOGIES: INDUSTRIAL BIOTECHNOLOGY IN AGRI/FOOD

Yvonne C. Armitage and A. Bryan Hanley
Knowledge Transfer Network, London, UK

INTRODUCTION

Biotechnology has been an important part of the food industry from its first inception. The first processes used in food processing, were cooking, fermentation and preservation and all three have been implicated in the evolution of human society.

The first great advance was the use of heat and the development of cooking. Processing of food by cooking increases the digestibility of food and this permits the development of larger brain sizes. This gives a distinct advantage to Homo sapiens compared to primates [1]. Indeed, typically only around 40% of the calories in uncooked food can be digested and therefore energy release is inefficient. Cooking remains a key component of both domestic and commercial food preparation. The development of new techniques such as microwave heating have dramatically changed cooking and processing practices and have enabled new types of food to be produced. In terms of industrial biotechnology, one of the key advances has been in the use of enzymes. The use of enzyme preparations such as thiol proteases (e.g., papain) to tenderise meat, chymosin and lipases in cheese making and pectinases in juice production have had a significant impact on subsequent food cooking and preparation. Some of examples will be given later in this chapter. These will help to illustrate the huge benefits that biotechnology have brought in the industrial processes and, subsequently, in the home setting.

A second major advance was the development of fermentation. Fermentation was used as both a way of creating new flavours and tastes and of preserving foods. Fermentation allows for the preservation of foods through control of degradation. Fermented drinks were being produced over 7,000 years ago in Babylon. Bread making was known to be common in Egypt over 3,500 years ago. The earliest evidence for fermented vegetables dates from over 3,000 years ago. In this case Aspergillus and Rhizopus moulds were most likely used and their use was documented in a book called Shu-Ching which was written in the Chou Dynasty (1121 – 256 BC) and which refers to the use of 'Chu' – a fermented grain [2]. The most common fermented food product was alcohol and wine casks have been dated to over 7,000 years ago. Lactic acid fermentation preserves food since the major product, lactic acid, is antimicrobial. The antimicrobial activity of lactic acid against nine bacteria (Escherichia coli, Proteus mirabilis, Salmonella enteritidis, Pseudomonas aeruginosa, Staphylococcus aureus, Enterococcus faecalis, Listeria monocytogenes, Bacillus cereus and Bacillus megaterium) and three yeasts (Rhodotorula sp., Saccharomyces cerevisiae and Candida albicans) was recently demonstrated [3] and the potential for the use of lactic acid in food systems was suggested. In addition to preservation, some strains of bacteria (particularly Lactobacillus and Bifidobacteria) have been implicated in gut health and, furthermore, in systemic health [4]. The presence of 'probiotic' bacteria has been suggested to be essential to maintaining a healthy internal microbial ecology and to prevention of dysbiosis. The human gut is a fermenter and, as with any such system, contamination in the form of unwanted microbial species can disturb the beneficial equilibrium and lead to a range of unwanted endpoints.

The third great advance, which is linked to fermentation, is the development of preservation techniques. Public health concern with food safety and food poisoning emerged in the UK in the latter part of the 19th century when a link began to be understood between food poisoning and specific bacterial species. Much of the interest prior to that was concerned with adulteration of food rather than microbial contamination however as incidences began to be recorded, the scale of the problem was further appreciated. Formal notification of food poisoning did not start in the UK until 1939 so, before that date, figures are relatively unreliable. However it is clear from the instances of cholera and diarrhoea that food borne infection was common. Heat treatment, canning and freezing all became popular as technology advanced and this had the dual benefit of improving diets, which were no longer governed strictly by seasonal availability of produce, and reducing the incidences of food borne microbial disease.

From these early beginnings, processing and biotechnology advanced to cope with ever increasing populations, changes in agricultural practice and the development of new ways of living. The development of new processes and new agri/food practices have grown together hand in hand and have been responsible for much of the progress of the past few centuries.

RECENT TECHNOLOGIES

A range of technologies have impacted upon the agri/food industry and, therefore, on consumers. The progression from small scale agricultural production to large scale and the increase in the processing capability of factories has been largely a consequence of industrial biotechnology. Many of the advances required chemistry as an enabling science.

AGRONOMY AND PRIMARY PRODUCTION

The techniques commonly used and developed in industrial biotechnology (gene mapping, metabolic profiling, new bio-based production methods) have had a significant impact all the way from primary production to final product. Indeed, such advances are essential if we are to overcome some of the major perturbations that threaten existing and future food production systems.

Agronomy and an increased understanding of plant metabolic pathways has led to greatly increased crop yields. In some instances it has been largely a question of trial and observation. For example, the consequences of magnesium deficiency are known and lead to yellowing and chlorosis, particularly of old leaves. The practical solution is to add magnesium. Magnesium deficiency, however, has similar physiological appearances to other deficiencies. On a molecular level, magnesium is essential for photosynthesis since it is at the centre of the corrin ring that is a key component of the photosynthetic pathway. By combining chemical analysis of soil using a multi-ion sensor (http://www.cleangrow.com/product/nutrient-analyzer) and plant tissue analysisit is possible to make a rapid and definitive diagnosis and take appropriate remedial action [5]. In this case, chemistry and an understanding of the basis of the pathology of a plant disease allows remedial action to be taken.

Plant Breeding

Traditional plant breeding and improvements in agronomic practice has led to significant increases in crop yield. According to FAO data, in less than 50 years (1961–2009) the world average cereal yield increased from 1.35 to 3.51 t/ha. The total production of cereals has almost trebled in this period with just an increase of just 9% in acreage under cultivation. In the U.S. yields of maize have increased fivefold since the 1930s. There are still substantial challenges to be faced. While increases in yield are possible without substantial increases in acreage, this means that owning the technology to convert plants into high value protein products will be increasingly important. In addition the use of bio-based feedstocks will continue to compete with food production. The development of new methods to increase the yields of crops, their resistance to disease and the more efficient use of waste materials (repurposing waste streams), will all require significant industrial biotechnology inputs.

ANIMAL PROTEIN FOR HUMAN USE

Growing populations and increasing wealth in Asia and other markets is leading to an increased demand for animal protein. The WHO have estimated that per capita consumption of animal protein will increase globally by 50% by 2030 and by >80% in Asia. Others have forecast an annual growth in consumption of >10% in Asia compared to 5% globally. Asia cannot meet such demand from internal resources [6]. By 2050 the global population is expected to rise from 7.4 billion today to 9.6 billion and will require 70 per cent more protein than is currently available. In the US (a primarily meat-based consumption pattern) the food production system uses about 50% of the total US land area, approximately 80% of the fresh water, and 17% of the fossil energy used in the country. Globally, livestock systems occupy about 30 per cent of the planet's ice-free terrestrial surface area. If, as noted above, a 70% increase in protein production will be necessary by 2050, it is hard to see how this could be met by increases in animal protein production alone without substantial increases in acreage devoted to animal production. Over the past 50 years, global meat production has almost quadrupled from 78 million tonnes in 1963 to a current total of 308 million tonnes per year. However per capita, meat consumption in China has increased six-fold over the past 40 years. Since the population almost doubled to 1.3 billion people over the same period, global demand for meat and animal feed has exploded. This pattern is likely to be repeated with even faster rates of growth in consumption. The use of genetic modification and the development of new protein sources (in some cases from single carbon feedstocks) will be an essential part of the drive to meet this challenge.

Food Security, including the provision of animal protein, is a key policy driver for the UK. Food security has many dimensions and responsibility for it rests with at least two departments—Defra, and BEIS. As the report of the House of Commons Environment, Food and Rural Affairs Committee on Food Security noted in 2014, the pressure on animal protein will have a knock-on effect not only on animal production itself, but on crops used for animal feed – particularly soya. The report concluded "We are concerned about the potential impact of projected rising trends in global demand for animal protein on the price of animal feeds and the cost of production."

So, in addition to human requirements, alternative sources of protein are needed for animal feed. There is a substantial world market, both in terrestrial livestock (and particularly pigs and poultry), and also aquaculture. Animal feed is a huge global market, with global commercial feed manufacturing generating an estimated annual turnover of over US $400 billion (international feed industry federation).

With livestock animal feed, there is a substantial commercial opportunity to provide an alternative to imported soya in diets. This presents a significant challenge for our domestic livestock industry, which accounts for more than 50% of the total UK agricultural output, but is less than 30% self sufficient in vegetable protein. Replacement proteins are needed due to ever growing concerns over the environmental sustainability of increasing production of South American soya, which accounts for over 90% of EU protein imports [7].

Aquaculture is also developing rapidly around the world, and now provides more fish than wild-catch fisheries. Fish are fed a protein rich diet, and there is a strong industry need for novel sources of vegetable-based proteins to replace fish-meal in diets.

While insect protein has been suggested to be a potential solution, ADAS have concluded "At present, insects cannot be fed to livestock in the European Union, following the ban on all animal protein in feed in 2001." This may be a potential block to this particular solution. In addition, the efficiency of production of protein has been suggested to be no better than that of chickens [8] although better than that of beef. The EU have recently changed the regulations concerning the feeding of insect protein to fish however feeding insects protein to other livestock is still banned and the ban is likely to stay in place for the foreseeable future for ruminant animals. Another change in the regulatory landscape will take effect on the 1st January 2018. The new Novel Foods Regulation (2015/2283) will for the first time, provide a clear route to approval of insects for human consumption via the European Food Standards Agency. The application procedure will be generic so once a given product is approved, other companies will not need to seek approval for the same product.

As the increasing population and global demand for more meat has become reality in the last few decades there has been significant research and innovation activity put towards improving meat productivity through animal breeding and nutrition. More recently, trends towards investigating alternative healthier mostly meat-free forms of protein has seen a rise in the number of vegetal/plant derived products such as soya, tofu together with well known brands such as the mycoprotein Quorn. However, increasingly there is need to develop a range of cost-effective products with different properties and composition particularly to meet different cultural and dietary needs. Thus, there is huge scope for innovation to develop new products at targeted demographic groups, particularly the elderly in order to support healthy aging. This will necessitate the development of new approaches to animal protein production and new technologies to grow, harvest and process novel sources of protein including plants, algae, yeast and bacterially derived materials.

Crop Protection

One of the substantial breakthroughs in technology came with the development of the means to control pests and diseases of plants and animals. While pesticides have been used for thousands of years, it is only relatively recently that science for the creation of synthetic pest control agents has been developed. Sulfur was used as a fumigant thousands of years ago and lead arsenate was first used in 1892 as an orchard spray. However, pests and disease still led to significant losses. One of the most infamous was the destruction of Ireland's potato crop by a pest in the nineteenth century. A plant disease called late blight ruined the potato crop, the staple food crop in Ireland at that time. Up to a million Europeans starved to death during the Great Irish Famine of 1845-1847. Social implications of this pest infestation in Ireland included the largest ever migration to the United States, with more than a million people migrating from Ireland during that famine.

The development of more effective control methods has greatly improved crop yields and the stability of the food supply. For example, the synthetic pyrethroid insecticides were developed at Rothamsted Research, in the 1960s and 1970s. Today they account for around one sixth of global insecticide sales and global annual sales of pyrethroids exceed US$1.4Bn. They are also used to impregnate bed nets, which help to reduce the spread of malaria as part of the World Health Organisation's Global Malaria Programme.

There have, however, been questions raised about the long-term effects of pesticide use. The WHO have classified pesticides according to their potential to cause acute harm to non-target species (humans).

Hazard Class: based on oral or dermal LD50 in mg/kg intervals
Class I-a: Extremely hazardous: <5 oral, <50 dermal
Class I-b: Highly hazardous: 5 - 50 oral, 50 - 200 dermal
Class II: Moderately hazardous: 50 - 2000 oral, 200 - 2000 dermal
Class III: Slightly hazardous: >2000 oral, >2000 dermal
Class U: Unlikely to present acute hazard >5000

Concerns about chronic exposure remain and there is growing pressure to limit the use of pesticides. One consequence of this has been the increase in science and technology around biological control methods. These methods are not always effective and can lead to the replacement of one nuisance with one that is even more deleterious. One often cited example is that of Cane Toads. The toads were introduced into Australia in 1935 from the Amazon as an experiment to control a beetle that devastated sugar cane crops. The initial 102 young toads quickly multiplied until there were hundreds of millions. The population can rise to a density of 2000 toads per hectare. They were ineffective as a pest control method and, even worse, because of their toxicity, they also depleted natural species – some of which were insect consumers. Cane toads were not only ineffective, they actually exacerbated the problem.

An alternative, more controllable and more biotechnologically based approach, is to use genetic modification (GM). Crops that have been genetically engineered to produce proteins from the bacterium Bacillus thuringiensis (Bt) have been used to control insect pests since 1996. However, while this is a less impactful approach compared to the use of uncontrolled biological methods, there has still been considerable resistance to the whole idea of using GM. In addition, some pests have become resistant to Bt crops that make only one toxin and therefore biotech companies introduced Bt crops called 'pyramids' that produce two or more Bt toxins active against the same pest. These have met with more success. A more specific approach is to make use of insect physiology and to target pest species only. Both spider and scorpion venoms have been used as a source of insect-specific neurotoxins [9]. While screening libraries of peptide toxins produced by a range of spider species has met with some success, a more targeted approach focused on understanding insect neurophysiology is also being adopted. Such approaches are aimed at finding analogues to insect neuropeptide hormones that are specific to crop plant consuming target species.

Molecular and Related Techniques

Many of the molecular techniques that have been adopted have specifically targeted pasts and diseases (e.g., prevention of pest damage and herbicide resistance). A number of crops are now able to be produced with enhanced nutritional or functional profiles. These include soya with an improved fatty acid profile, potatoes that are resistant to bruising, virus resistant papaya, non-browning apples and alfalfa with reduced levels of lignin to improve digestibility. There are a number of other targets including plants with enhanced drought tolerance, better micronutrient profiles (e.g., calcium or certain vitamins), improved absorption (e.g., iron), more beneficial antioxidants (e.g., anthocyanins), vaccines and biopharmaceuticals [10]. The use of genetic modification to directly enhance crops and crop quality has the potential to have a significant impact on the agri food industry.

In addition to plant protection, a better understanding of biosynthesis and a range of ancillary techniques including next generation sequencing has allowed science to create designed crops and, potentially, animals. The use of trait markers and techniques for gene editing such as CRISPR/Cas9 system has great potential to improve crop plants. CRISPR stands for Clustered Regularly Interspaced Short Palindromic Repeats and Cas9 is CRISPR associated protein-9 nuclease. The way the system works is that Cas9 can cut DNA at a specific location. A synthetic guide RNA is used to direct the nuclease to introduce a double strand break at a specific location within a strand of DNA. When the cell repairs that break, errors can occur to generate a gene knockout or additional genetic modifications can be introduced. This type of gene editing has potential for the creation of species with specifically modified and improved characteristics.

The use of molecular techniques to create food with defined characteristics could be of great value in being able to make health claims about specific foods. One of the major drawbacks to the development of health claims is the inconsistency of the raw materials that are used and the presence of bioactive confounders. This may be overcome using cloning and gene editing.

Despite the potential and promise of this type of advanced molecular science, it will be necessary to ensure that chemistry keeps pace with biological advances through the adoption and development of analytical fingerprinting techniques and that people have confidence in the science and understand what they are being asked to consume.

Food Processing

Moving further along the food chain, advances in processing have led to significant improvements in how food is produced. A number of these are of historical importance however many are still used and are of significant commercial value. For the sake of brevity, we will concentrate on those specifically related to biotechnological advances in the food processing area.

Flavour Production

While we have mentioned the use of industrial biotechnology and related techniques for the production of food staples (particularly protein), one of the most significant areas of current (and likely future) use of industrial biotechnology in food will be in the production of high value components. Principal among these is the production of flavours.

Natural flavours attract a premium price therefore nature-identical options are popular and most of these are produced through the use of industrial biotechnology.

While the use of flavour extracts and tinctures was common practice as early as the 13th century, many of these were prepared for their supposed medicinal properties. The first purely synthetic flavours included esters such as methyl salicylate (1859, wintergreen oil), aromatic compounds such as benzaldehyde (1870, artificial butter oil) and vanillin (1872). A range of esters were then prepared to mimic fruit flavors. These included 3-methylbutyl acetate (banana), ethyl acetate (general aroma enhancer), pentyl acetate (pears), octyl acetate (oranges), methylpropyl formate (raspberry), methyl butyrate (apple) and methyl anthranilate (grape). Many of these compounds also occur naturally in the fruit in question however the flavour notes are often much more complex than those of the esters. There are two other significant challenges for the flavour industry – structurally different compounds having the same or similar taste characteristics and chirality.

There a number of examples of compounds which have a similar taste profile but which are structurally very different. A simple example of this are the range of very structurally diverse natural and synthetic sweeteners. While their taste profiles are not identical, they contain enough communality to suggest that there ought to at least some structural characteristics in common however they range from classic sugars to dipeptides (aspartame) and saccharin which is a benzothiazolone.

Chirality is important in taste molecules. One of the classic examples, which was established in the 1970s, is the case of carvone (2-methyl-5-(prop-1-en-2-yl) cyclohex-2-en-1-one). One enantiomer (the R – form) is the main constituent of spearmint while the other (the S + form) is responsible for the smell and taste of caraway seeds.

It is clear that from a value and flavour standpoint, the use of natural flavours is greatly favoured. These are expensive and rely on there being adequate natural sources and a consistent supply. It is this gap between authenticity and practicality, consistency of supply and cost that industrial biotechnology is able to offer an alternative through the use of biological processes to create nature identical flavours (chiral where necessary) at a lower cost and with a consistency of flavour profile.

Additive Manufacturing

The use of additive manufacturing in food – more commonly referred to as 3D printing – is becoming increasingly talked about. However, it is unclear that there is any particular value beyond the creation of interesting shapes and some novel combinations of ingredients. The challenge for food scientists, technologists and producers is to find propositions with added value that are attractive to consumers at a cost that reflects that added value and which are not able to be produced in other ways.

In practical and simple terms, 3D printing is a way of creating an object from a 3D scan by building up successive layers of the constituents of the product until the complete object is created. According to the American Society for Testing and Materials Group, there are various techniques that can be used to build up the layers and these include Material Jetting, Binder Jetting, Material Extrusion and others (http://www.astm.org/COMMITTEE/F42.htm). These are essentially industrial processes that are more usually applied to stable feedstocks such as plastics and the slow application of 3D printing of foods is in part at least, a consequence of the unfamiliarity of food technologists with the types of processes that are applied in the 3D printing sphere in other sectors.

A recently funded EU project on personalised nutrition for the elderly [11] aimed to produce nutritionally adequate, complete meals for the elderly (who may have mastication and swallowing difficulties) using 3D printing. The focus of the project was on the older population but other demographics could be targeted. A similar approach has been suggested to make edible insects more acceptable by using 3D printing to remove visual insect characteristics and to create versions of current foods.

It should be possible to create food manufacturing systems which are essentially 'feedstock blind' and which can produce foods from a specified protein, carbohydrate, fat and sugar mixture. This may be of increased value as we move towards potential shortages of conventional crops. The use of advanced biotechnology to create foods designed to have specific properties shows promise but there has been little realisation of that promise thus far.

Meat Production

The concept of meat grown in labs has been around for a considerable time. Artificial meat has been produced from a fungal source for a number of years and the product (Quorn) has proved to be increasingly popular. The original work by ICI led to production of Pruteen – a single cell protein produced by Methophylus methylotropus from methanol which was proposed as both a food and feed. Vegetarian meat substitutes including those grown from pulses such as soya and peas have been a part of the normal diet for many years. NASA had a significant programme of work associated with finding alternatives protein sources for long space trips and experimented with algae.

However, the idea that it might be possible to produce a truly synthetic meat in a test tube is a fairly recent innovation. In principle, culturing animal muscle cells so as to make a meat product such as a hamburger has been demonstrated. The drawback is that the only burger produced so far cost $300,000 to produce and took over two years. The idea of affordable synthetic meat continues to attract investment and two companies Memphis Meats and Mosa Meats hope to have commercially available products ready by 2020. There are a number of hurdles to overcome including the requirements for the cells to have a medium (generally serum derived) to grow in and the fact that meat also contains fat and this contributes to the characteristics of the final product. The term 'clean meat' has begun to be used by organisations such as PETA who are showing some interst in a 'cruelty free' synthetic meat substitute. In addition, from a sustainability standpoint, laboratory produced meat uses 10% of the water and 1% of the land required for conventional meat production. There are a number of societal and regulatory hurdles that will have to be overcome however the main consumer drivers are still likely to be cost and taste. Experience suggests that technology is likely to be able to produce new and interesting foods and feeds.

CONCLUSION

Biotechnology has been at the core of advances in agriculture and the conversion of raw materials into food. As a result of advances in the underpinning science and the practical application of that science, we now have a range of techniques that are available. The increasing minimisation of waste streams and their use to create nutritious food from previously unused sources has led to significant advances in diet and nutrition. Clearly the population increase and heightened expectations referred to in other chapters will require ever greater levels of ingenuity and imagination.

The development of alternative food crops and the increased realisation of the importance of minor components to health will spur further research and the development of new and more efficient technologies.

While new and novel technologies are important, it is worth remembering that there are also still classic IB approaches that meet a significant need. For example, the largest enzyme-mediated reaction that exists is the production of high fructose corn syrup through the action of glucose isomerase on glucose. Another common food manufacturing process also relies on a by-product of a biotechnological process – the production of Marmite as a result of autolysis of yeast from the brewing industry.

Industrial biotechnology will continue to be at the centre of food and food ingredient production – from lower value bulk materials all the way to the small and valuable flavours, colours and fragrances that are so essential to product development.

References

[1] *The Human Advantage: A New Understanding of How our Brains Became Remarkable. Suzana Herculano-Houzel*, MIT Press, 2016.

[2] Fermented Fruits and Vegetables: a Global Perspective, *FAO Agricultural Services Bulletin No. 134*, 1998.

[3] Stanojević-Nikolić, S., Dimić, G., Mojović, L., Pejin, J., Djukić-Vuković, A. and Kocić-Tanackov, S. (2016), Antimicrobial Activity of Lactic Acid Against Pathogen and Spoilage Microorganisms. *Journal of Food Processing and Preservation*, 40: 990–998. doi:10.1111/jfpp.12679.

[4] Impact of the gut microbiota on inflammation, obesity, and metabolic disease. Boulangé, Claire L., Neves, Ana Luisa, Chilloux, Julien, Nicholson Jeremy K. and Dumas, Marc-Emmanuel. *Genome Medicine* 20168:42.

[5] Limantara et al., *Procedia Chemistry* Volume 14, 2015, Pages 225-231.

[6] Wu et al., *Ann N Y Acad Sci.* 2014 Aug;1321: 1-19.

[7] Fefac – the European Feed Manufacturers' Federation, *Statistical Yearbook,* 2009.

[8] http://journals.plos.org/plosone/article?id=10.1371/journal.pone.0118785.

[9] Windley, Monique J., Escoubas, Pierre, Valenzuela Stella M. and Nicholson, Graham M. *Molecular Pharmacology* July 2011, 80 (1) 1-13.

[10] Hefferon and Choi, *Int J Mol Sci*, 16, 3895, 2015.
[11] http://www.performance-fp7.eu.

In: Food: The Vital Ingredient ISBN: 978-1-53614-402-4
Editors: A. B. Hanley and W. R. Russell

Chapter 12

THE CHEMISTRY AND PHYSICS OF EATING

Peter Lillford
University of Birmingham, Birmingham, UK

INTRODUCTION

Food is only nutritious if it is eaten, and every consumer will readily report which foods they like to eat and those they do not. These judgements are conditioned by the culture of the cuisine they are used to, their age, their circumstances, the time of day etc. Preferences can be learned or taught but most of the measurements supporting liking are conducted by the eyes, nose and within the mouth, which are measurement devices far more sophisticated than most in- vitro sensors in the laboratory.

With the exception of fresh whole fruit and raw milk, most foods are processed in the home, by a chef, or in a factory before eating. For some materials this processing is minimal such as the cutting and cooking of meat fish and vegetables ("Natural Foods"), whereas for others, the raw materials are rebuilt into wholly new structures (Baked cereal goods, snacks, cheese, confectionery). These can be considered as the class of "Fabricated Foods" where, without the processor, the raw materials would be difficult to eat.

Food processing has developed empirically over millennia, and the processes and raw materials are combined to form something that we expect to enjoy. There are many other combinations which can be performed which produce a food that we do not enjoy or cannot eat, but these have not survived. So, selection for consumer acceptance has already happened. We now live in an era dependent on mass produced food (see Chapter 8), but it is important to recognise that mass manufacture has chosen foods already liked, and by applying economies of scale produced them with greater energy efficiency, lower wastage and therefore reduced cost. Engineering therefore was the dominant discipline as scales increased, helping to transform factories from batch processes in "big kitchens" to continuous coupled lines with minimal downtime and maximised throughput. This approach of optimising throughput of acceptable products leads to rigid raw material, the use of ingredients which are "process aids", and inflexible processes. ("If it works, don't fix it"). Provided quality assurance measurements are in place, this standardisation means that success is almost guaranteed. However, this approach has added little to our understanding of why we like the food provided, so we know very little about the chemical and physical stimuli that induce liking, even in commonplace foods. Despite much research in recent years, this gap in knowledge is broadening because of the need to reformulate foods which is driven by fluctuations and scarcity of traditional raw materials (Climate Change); a drive to use materials more locally (Food Miles;) the need to reduce calorific density (Obesity and Metabolic Syndrome); and the growing requirement of consumers for health benefits (Superfoods) including elimination of chemically derived process aids. (Clean Label).

There is no doubt that the culinary skills of a well-trained chef will allow innovation against all of these targets. However, chefs are also empirical experimentalists, usually expert in a particular food culture. They will become vital partners in the demonstration of acceptable foods from novel materials, but for new supply chains to grow efficiently, scale up will be necessary from these artisanal skills to efficient manufacture. Now we must design foods for pleasurable eating with novel materials. What are the stimuli for a response leading to pleasurable eating?

Measurement of Sensory Response

As well as quantifying consumer responses to liking (Hedonics), the consumer can be trained to score the sensory attributes of a food on an agreed measurement scale, using a lexicon of descriptors with defined values. This uses a cohort of individuals as a measurement tool, reporting perception rather than liking. The information gathered is a response to multiple stimuli, usually related to a particular food type. The degree of resolution is determined by the selection of the lexicon, and the accuracy of output relates to the degree of training and the selection of panellists. Pioneering work was carried out by Szczesniak [1] who identified responses to Appearance, Flavour, Taste, and Texture. Her group also attempted to couple each of the measured responses to physical or chemical measures which provided some clue to the stimulus causing it. [2].

Most of the research since then has focussed on the better definition of responses and more sophisticated analysis of the stimuli. This has led to the recognition that multidisciplinary studies are required, using (bio)chemistry, physics, engineering, human physiology and neuroscience, and behavioural psychology [3].

In addition, we have learned that stimuli are presented as a function of time, so that responses also reflect a dynamic process where sensations are detected sequentially. This leads to measurement of the temporal dominance of responses, rather than a simple list of attributes [4]. Furthermore, liking is related to the alignment of a food with its expected performance. Any deviation can trigger a dislike. In other words, humans interact with foods in a scripted manner, where deviations from expectation are dominant in our perception and liking [5]. This explains much of the effects of culture, age and experience in our acceptance and liking of food types.

Measurement of Stimuli

Appearance

Our first interaction with food is when we see it. This does not just refer to colour and turbidity, but also to overall appearance (uniformity of size and shape), and surface appearance of gloss, visible structure (heterogeneity, fibrosity etc.). Probably because the function of the eye is well understood, and the properties relate to the undisturbed properties of the food, there is a good understanding of how to measure the stimuli which cause the sensory response [6].

The enormous power of computer graphics now allows "virtual" experiments to be done where images can be easily adjusted in size, surface texture, shape, colour etc, and the resultant sensory response obtained [7].

Flavour

Flavour can be regarded as a response to two types of stimuli; volatile odours taken directly into nose, or retronasally during chewing and swallowing; and taste, the stimulation of receptors on the tongue and throughout the buccal cavity [8].

Odour

The detection of volatiles is primarily via their binding to receptor proteins in the epithelia of the nasal cavity [9]. The human genome contains 387 genes capable of expressing receptor proteins. Each protein is not exclusive to a particular odorant, but classes of proteins exhibit preferential binding of a particular functional groups within odorants [10].

Given this complexity, it is not surprising that a detailed mechanism for the threshold and intensity of single molecular stimulants is not yet known, and therefore the composition required to deliver a particular flavour profile (such as apple, strawberry or butter) cannot be predicted. Instead, the current focus is on measuring the composition of flavour types which can now be done in great detail, through the use of GC -Mass spectroscopy, high resolution NMR etc. (See also Chapter 5). Even with such powerful qualitative and quantitative tools, reconstitution of the mixture does not exactly match the sensory properties measured by trained panels [11].

This is not too surprising when we consider the complex chemistry and structure of most foods. Aroma molecules have different solubility in the solution phases of water and oil, Since the aromas are detected as volatiles, their proportions in the headspace above the food are determined by their partition coefficients in any phase systems present. Secondly, because most foods are not at chemical equilibrium, but contain kinetically trapped phases, the kinetics of release during mastication determines the rates of delivery of each odorant, and therefore the time dependence of their stimulus and response. For liquid foods, measurement of the dynamic headspace above a stirred system gives a better representation of how stimuli are delivered. For solid and semi solid foods, sampling from the nose of a human subject, or the use of a "model mouth" in which both mastication and mixing is simulated, is a necessary system [12]. It is not yet possible to model and predict the dynamics of aroma release from any food, because the necessary data is not available. Now that the consumer requires reduced or lower fat foods with matched odour release, experimental measurement rather than prediction of formulation will remain for many years to come.

Taste

Four key tastes were originally specified as Sweet, Sour, Bitter and Salty, now other tastes such as savoury (Umami) and possibly Fatty have been added. Also, the trigeminal effect of heating and cooling have been recognised for some time. The detection of tastes is primarily via receptors present in the taste buds of papillae in the tongue, but receptors have also been identified elsewhere in the digestive tract and lungs.

The physiological mechanisms of action have been studied extensively in recent years, and are complex, for example: -Trigeminal receptor TRPV1 is identified with the heat of capsaicin, whereas CMR1 detects cooling by menthol and eucalyptus. Salty and sour taste sensations are both detected through ion channels. Sweet, bitter, and umami tastes are detected by G protein-coupled taste receptors. (An excellent summary of the physiology, neurology and genetics of receptors can be found in “Taste Receptors”, Wikipedia). Though the application of molecular biology and animal models has rapidly advanced in recent years, there is still much which is unknown. It is clear however, that the distribution and therefore the stimulation of taste buds can vary dramatically between human subjects. This explains the identification of “Supertasters” amongst food panellists, and reinforces the need for high numbers in panels before significant, averaged differences between food products can be properly identified. Work in this field concentrates on the single stimuli in a common delivery food type. The comments above concerning the dynamics of release during mastication of whole foods also applies with tastes, so time dependence of tastes can be assayed, and the structural breakdown of the food itself effects the timing and intensity of stimuli.

This understanding has been used in techniques for salt reduction, together with other combinations of tastants [13, 14].

Texture

Bourne (15) defined the texture of food as:

1) A group of physical properties that derive from the structure of the food.
2) It belongs under the mechanical or rheological subheading of physical properties.
3) It is sensed primarily by the feeling of touch.

The mechanoreceptors in the oral cavity appear to be of 3 types; Slow Acting SA1, which are sensitive to the form and structure of particles; Slow acting SA2, which sense the position and action of the tongue, in collecting large particles and manipulating the bolus prior to swallowing; and Fast acting FA1, whose action in the mouth may relate to vibrational skin stretching during slip. The action of mechanoreceptors in the mouth is far less well understood that on external skin, since the usual method of study is by touch with microprobes, and mapping by measurement of the subsequent neural response (microneurography). A more detailed review of taste and nociceptors can be found in "Food Oral Processing" [16].

Hence there are almost no studies of real foods, but only of simple models of particles in suspending fluids. Instead, a different approach has been taken, where the physical properties of foods are measured, and correlated with sensory attribute perception. Where general principles emerge, causal models between physical properties (potential stimuli) and sensory response are proposed and tested. The historical development of texture studies has been reviewed recently [17]. For liquids, the measurement of shear stress and strain rates (elements of viscosity) showed that for low viscosity drinks, the mouth is apparently measuring flow rates (shear rates), whereas for higher viscosity foods, requiring little or no mastication, the shear stress (squeezing force) was the sensory stimulus. More recently, tribological measures have been applied [18], but the presence of saliva is crucial in obtaining relevant measurements, so the science is still emerging.

For solid foods, the microstructure of products is intrinsic to their texture perception, and the complete breakdown path from whole product to swallowable bolus needs to be examined. Early work correlated physical testing with sensory properties. Tests were developed to mimic mouth action, but their complex geometries do not permit the identification of real stress and strain during deformation and fracture. Nonetheless they remain a good indicator of first bite properties such as hardness, brittleness etc., and remain as a quality control method. Later work has considered fracture and failure mechanics of composite foods, borrowing the measurement methods and theoretical models from composite physics. Since foods present a variety of fibrous, aligned and aerated structures, breakdown is best established by treating the mouth as a comminution machine, sampling structures during the chewing sequence, and examining their microscopic structures. Then, the relevant mechanical tests, related to the in- mouth process can be applied [19].

There is still a major uncertainty in determining the real in-mouth stimuli late in the breakdown path. Saliva is stimulated, and contributes to lubrication, and reassembly to a swallowable bolus. In vitro measurements are difficult and unreliable, due to the instability and molecular variability of saliva components. Mechanical studies on expectorated foods should shed light on the subject but are very rare. However, the ageing population displays increasing number of elderly consumers with different levels of dysphagia, which a limit on food choice and influences nutritional status. Foods now require reformulation simply to allow swallowing, but with the maintenance of pleasurable texture sensations.

SUMMARY

This paper has considered the separate food attributes of Appearance, Flavour and Texture, and our state of knowledge concerning their stimulus. But this separation is itself artificial. We all know that the appearance of foods can stimulate salivation, as can the presence of certain tastants and aromas. This in turn will modulate their breakdown during mastication and perceived texture. We are far from building a unified model, let alone predicting the hedonic qualities of food from formulation and process alone. Further development will require multidisciplinary engagement by chemists, material scientists and engineers in creating foods, whilst biologist and neuroscientists develop their understanding of how stimuli are recognised and converted via neural signals to the sensations we enjoy. All this, whilst increasing demand for higher sensory pleasure and appropriate nutritional balance are challenged by limiting raw material supply and the need for an increase in the sustainability of practices throughout the food chain.

REFERENCES

[1] Szczesniak, A. S., & Kleyn, D. H. (1963). "Consumer awareness of texture and other food attributes." *Food Technology*, 17, 74–77.

[2] Brandt, M. A., Skinner, L. Z., & Coleman, J. A. (1963). "Texture profile method." *Journal of Food Science*, 28, 404–409.

[3] Aguilera, J. M., (2018)," Food Engineering into the XXI Century", *AIChE Journal*, 18, 64, 2-11.

[4] Pineaua, N. et al., (2009), "Temporal Dominance of Sensations: Construction of the TDS curves and comparison with time–intensity", *Food Quality and Preference*, 20, 450–455.

[5] Erasmus, A. C., Boshoff, E. and Rousseau, G. G. (2002), "The potential of using script theory in consumer behaviour research", *Journal of Family Ecology and Consumer Sciences*, 30, 1-9.

[6] Hutchings, J. B., (1999) "*Food Colour and Appearance*", Springer US.

[7] Hutchings et al., (2013), "Food Appearance, Quality Assessment and Specification" in "I*nstrumental Assessment of Food Quality*", Woodhead Publishing, pp. 29-52.

[8] Laing, D. J. and Jinks, A., (1996), "Flavour Perception Mechanism", *Trends in Food Science & Technology*, 7,_12, 387-389.

[9] Buck, L. and Axel, R., (1991) "A novel multigene family may encode odorant receptors: a molecular basis for odour recognition." *Cell* 65, 1, 175–87.

[10] Niimura, Y., Nei, M. (2007) "Extensive gains and losses of olfactory receptor genes in mammalian evolution." *PLoS One*. 2 (8): 708-713.

[11] Peterson, D. G. and Reineccius, G. A., (2003) "Characterization of the volatile compounds that constitute fresh sweet cream butter aroma", *Flavour Fragr. J.*, 18, 215–220.

[12] Taylor, A. J., and Linforth, R. S. (1996) "Flavour release in the mouth", *Trends in Food Science & Technology*, 7, 12, 444-448).

[13] Toldrá, F. and Barat, J. M. (2009), "Recent patents for sodium reduction in foods." *Recent Pat. Food Nutr. Agric.* 1, 80-86.

[14] Liem, D. G. Miremadi F. and Keast R. S., "*Reducing Sodium in Foods: The Effect on Flavour*" Nutrients, www.mdpi.com/journal/nutrients).

[15] Bourne, M. (2004). "Relation between texture and mastication". *Journal of Texture Studies*, 35, 125–143.

[16] Chen, J. and Engelen, L. (2012), "*Food Oral Processing: Fundamentals of Eating and Sensory Perception*", Wiley-Blackwell.

[17] Lillford, P. J. (2017), "Texture and breakdown in the mouth: An industrial research approach", *J. Texture Studies*. 1–6. DOI: 10.1111/jtxs.12279.

[18] Chen, J., and Stokes, J. R. (2012). Rheology and tribology: Two distinctive regimes of food texture sensation. *Trends in Food Science & Technology*, 25, 4–12.

[19] Lillford, P. J. (2001). Mechanisms of fracture in foods. *Journal of Texture Studies*, 32(5), 397–417.

CONCLUSION AND RECOMMENDATIONS

Chemistry is a key enabling science and remains at the forefront of the future of industry, research and future socioeconomic benefits. The food and drink sector is no exception. The chapters in this book have reviewed both opportunities and challenges. We live in a competitive environment – both globally and locally and chemistry must demonstrate its relevance and the rewards it will enable. This is a challenging but also exciting environment. Some of these key challenges are summarised below in the context of individuals, ideas, innovation, infrastructure and investment – the 5i approach.

Individuals - A Skilled Workforce; and More

Many parts of the food chain are considered to have a relatively low skills base. On the other hand, some of the most advanced and efficient manufacturing systems are also widely used. The development of science and technology, underpinned by research, is crucial to future progress. There is no point in developing new systems, smarter ways of working and new and innovative technologies if people feel that their livelihoods are threatened. That is why upskilling along the full food chain is a key prerequisite for the introduction of new and better technologies. This is particularly relevant with regard to innovation and technology transfer to low and medium income countries, some of whom lack the infrastructure considered necessary for advanced industrialisation. Everything from agronomic practice, harvesting, storage to primary and secondary processing and presentation of products to the consumer has to work in harmony. This means that individuals must have some appreciation, not only of their own role and job, but also how it will impact upon the rest of the food chain. This is a bidirectional requirement. If scientists and technologists are not able to translate the needs of consumers into product characteristics then producers will not be able to fulfil those needs. For example, if the health benefits of specific foods and food components are not conveyed to the individual consumer in a clear and concise manner, then products will fail and, perhaps more importantly, their potential benefits will not be realised. The need for 'upskilling' does not just apply to those who work in the agri-food industry. It also applies to those in ancillary industries (e.g., packaging, robotics and AI, health and biosciences). Furthermore consumers need to have an understanding of the products they are buying and their characteristics in order to be able to make informed choices.

There are some key requirements if the food and drink industry is to grow and continue to be successful.

- There needs to be a better flow of high level skills from academia and R&D into the food and drink industry. This will enable advances to be quickly incorporated into industrial practice and help to overcome the 'valley of death' syndrome where innovation fails to bridge the gap between research and successful application. There have been a number of schemes but these are invariably piecemeal in nature and there is a lack of coherence. The UK has an opportunity to adopt and to improve current best practice and thereby to spread this globally both in terms of the skills themselves and how these atr translated.
- There needs to be better defined career paths and a better system for the adoption of new skills as new technologies and new products

with novel characteristics (e.g., healthy ingredients) are developed in response to consumer need and social drivers.

- A national scheme along the lines of the National Retraining Scheme, which is part of the UK Industrial Strategy is a crucial component and active participation by industry and those working in the industry is essential. In an international context, the loss of access by the UK to EU schemes such as Marie Curie may have a significant impact on university sector training. The lack of international conformity of apprenticeships and other types of training will continue to act as a brake on innovation. New and more aligned systems would benefit both the sector and, in particular, companies with operations in several countries.
- In the UK the involvement of sectors and sector deals can provide a mechanism for the development of medium and long term aims for retraining and reskilling.
- There are a number of organisations that already have the means to reach out to people working in the agri-food industry. While many of these operate at a fairly advanced level (e.g., the Royal Society of Chemistry), there are opportunities for these and others (e.g., the Institute of Food Science and Technology) to be both at the forefront of knowledge creation, application and skills. In an international context, links between the RSC and its sister organisations in other countries may be of particular value. The use of the International Union of Pure and Applied Chemistry (IUPAC) as a facilitation organisation may be useful. The IFST has strong links to the Institute of Food Technology (IFT) in the US and both have a presence in other geographies.

Ideas - A Creative Future

The creative underpinning for the future of the agri-food sector will continue to be an absolute requirement. This will not only derive from food science but also from science in related disciplines. There are already a number of universities that offer Masters and Doctoral programmes in Food Science (a total of 107 separate programmes exist in the UK, currently). Some of these are specific to a part of the sector – for example, food safety, food security, food processing and production – while others are to be found in other specialisms such as nutrition, human physiology, plant sciences, physics, bioengineering and sports nutrition. While much of the research oriented effort is fragmentary, there are specific Doctoral Training Programmes in Agriculture and Food Security (University of Cambridge, SWBio DTP), Food Security (University of Reading, University of Nottingham) and others where agri-food is embedded in other programmes – for example, the N8 partnership. There are a number of universities that offer a Food Science degree. These include the universities of Surrey, Leeds, KCL, QUB, Reading, Nottingham, Newcastle, Lincoln, Ulster and Coventry. Fees for international students are typically £15 – 20,000 per annum. This compares with an average four year degree total cost in the USA of $150,000. In comparison, tuition fees in the Netherlands for EU or EEA students are about €2,000 per year. For students from outside these areas, fees are typically twice or three times that with a total degree cost of €24,000.

In a global context, the IFT lists approved undergraduate programs both in the US and elsewhere. There are a number of web sites that list organisations that provide Masters and doctoral degree programs. Some of these also provide information on potential sources of funding.

There are comparatively few sources of research funding specifically for food chemistry research. Some areas, for example, analytical chemistry, have clear applications in the food area while others are more focused on adjacent areas with relatively little cohesive input from food chemistry. In the US there is substantially more structure to food chemistry research however the focus has also been shifting away from food chemistry and towards other, related areas that have a higher profile and are considered to be more interesting.

It is essential that chemists working in the agri-food area continue to look outside their own discipline to find examples and opportunities to bring the clearest and most advanced thinking to the complex problems, challenges and opportunities in their area. More awards, prizes and other incentives are needed to provide further impetus for the best chemists to turn their thinking towards agri-food chemistry. More cross-disciplinary funding is needed where chemistry sits at the core rather than being an adjunct to other disciplines.

Every area of chemistry including synthetic chemistry, natural product research, QSAR, the chemistry of complex systems, quantum and theoretical chemistry, sensors, physicochemical reaction dynamics, measurement, rheology and the chemistry of biological systems can be applied to the agri-food area. The challenge is to create compelling and interesting problems that researchers can then publish in the most prestigious journals. Even a cursory glance at Nature Chemistry indicates that few food chemistry papers are submitted and accepted however the launch of a new journal by the Nature group (Science of Food) as an online and open access journal, will help to increase the impact and visibility of food chemistry.

Innovation – Chemistry at Work

Innovation has been described in a range of ways. However one of the most important aspects of innovation is that it goes beyond the creation of new ideas and new ways of thinking. The innovation process is more to do with the application of ideas rather than their creation. However in areas such as agri-food, a crucial and further step in the process of innovation is that the idea has to be accepted by customers since this will drive the adoption of the new technology that is a consequence of the new ideas.

There is are a number of consumer concerns about chemistry in the agri-food area. The Food Standards Agency commissioned a report to investigate consumer principles in the use of additives and enzymes. (Consumer Principles on the use of Food Additives and Enzymes, Kantor Public Research Report, 2017). Consumers made a strong association between 'unnatural' synthetic food additives and the potential for serious, long-term health risks. People also wanted to know about natural alternatives to synthetic chemicals in foods. Such ideas, often based on serious misconceptions, are exacerbated by some pressure groups who allow these misconceptions to flourish. The idea that natural is good and chemicals are bad is insidious and widely prevalent. The use of the precautionary principle in Europe should allow consumers to have greater levels of confidence in the food supply and, while there is considerable vocal opposition to food additives and processing aids, it is not clear that this actually affects consumer purchasing habits.

While such ideas do not, for obvious reasons, inhibit the development of pharmaceuticals, this is not the case for foods. Once ideas have become established, they are difficult to shift. For example, one of the most railed against food constituent is monosodium glutamate (MSG). Despite the fact that it occurs naturally in a range of foods, it is still widely condemned by those who think natural is best. Natural compounds are among the most toxic known with fungal toxins, ricin and goitrogens such as those found in Brassica vegetables having a range of toxic effects. One of the more illuminating set of concerns arose over acrylamide. Once upon a time, acrylamide was extensively used in the water treatment industry and for making polyacrylamide. It is subject to control in drinking water based on its toxicological profile. When it was found to be a natural component of a number of foods as a result of a Maillard reaction between asparagine and co-occurring reducing sugars, there was worldwide concern however it failed to register with many pressure groups.

Chemists must continue to make efforts to explain that chemicals are ubiquitous and that the use of clever and innovative chemistry has been, and will continue to be of great benefit to the planet and the people living on it. It is through the use of chemistry that we will be able to feed the growing global population and deal effectively with the many challenges that we will face in the next 50 years.

Infrastructure – Smart Building and Efficient Resource Utilisation

Infrastructure is essential to the effective growth of businesses and the economy. In addition to the digital infrastructure projects that have been identified by the UK government and which will be funded through initiatives such as the National Infrastructure and Construction Pipeline and the National Productivity Investment Fund. From an agri-food standpoint, initiatives to reduce waste, to cut carbon emissions, to broaden and improve transport links and to respond to global challenges will underpin growth. There has been a loss of significant industrial R&D facilities in the agri-food sector over the past 20 years and a growth in virtual centres and a recognition of a need to share knowledge. There is a need for new and more efficient facilities to process agricultural produce and to make food. This means that new, modern production facilities will have to be built alongside integrated research and development and, most importantly, appropriate scale up capabilities.

In addition to new factories and facilities, reform of infrastructure will enable the agri-food industry to respond to national and global challenges.

These include:

- The growth of a clean agri-food economy. According to UK government figures, the clean economy could grow at up to four times the rate of GDP. To underpin this growth, new chemistry will be at the core of the development of new and more energy efficient processes.
- New energy sources (wave and wind as well as solar) will transform how factories and businesses are powered. This will change energy supply lines. In order to achieve these gains, new chemistry will be needed to underpin energy creation and storage.
- Decreasing waste and improving efficiency will require new methods to be developed to maintain and use complex molecular structures (e.g., high value chemicals from waste). This in turn will need the development of new types of processors and reactors

underpinned by new chemistry. The UK government is committed to work with the food industry to deliver a 20% per capita reduction in food waste by 2025, through the Courtauld Commission. In addition to finding new uses for waste, it will also be necessary to decrease the amount of waste created. Chemistry will underpin new technologies and new processes in food preparation and packaging to enable this commitment to be met. The UN sustainable development goals recognise the importance of eliminating food waste and goal 12 calls for a reduction by half of per capita global food waste at the retail and consumer lrvrl by 2030 and reduction of food losses along production and supply chains.

As has been noted, raw materials, particularly animal protein and, to an extent, plant protein, will come under increasing pressure with a real danger of shortages as global populations and expectations increase. New methods for the production of protein and other feedstocks offers both a challenge and an opportunity. However, once produced, these must then be used to create food that is both healthy and wanted by the consumer. Understanding molecular and macromolecular interactions is essential to the design of new ways to create and use feedstocks and chemistry is at the centre of this requirement

These and other infrastructure requirements will need to be met if the agri-food industry is to meet the challenges and opportunities that will become an increasingly urgent aspect of our progress. Chemistry is essential not only in meeting the challenges but also in designing the infrastructure needs of the industry in the future.

Investment – A Vital Ingredient

Investment is a crucial part of the success of any sector and agri-food is no exception. Of the top 1000 R&D investors in the EU, a total of 26 were classified as food producers and their combined on R&D in 2015 was €2.7bn. This compares with some leading companies in other sectors including Volkswagen (€13.6bn) and Sanofi (€5.6bn). Volkswagen spent about 5.2% of revenue on R&D in 2013. This compares with Microsoft (13%) and Roche (19%). According to Food and Drink Europe, R&D spend by the food industry in Europe was around €3bn on a turnover of €1,244bn – about 0.25%

However R&D spend is not necessarily a reflection of innovation and inter-sectoral comparisons are difficult due to different definitions of what constitutes R&D. In addition, the food and drink sector is highly populated with SMEs who typically spend less on R&D than larger companies. Of the 289,000 food and drink companies in Europe, around 90% of them are SMEs.

Nonetheless the GVA of the food and drink industry is of the order of 1.8% in Europe. The figure for the UK (BEIS Report 2011) was around £70,000 per capita for the food and drink industry compared to £265,000 for pharmaceuticals however the overall figure for non-R&D utilising manufacturing was £37,000.

The UK government has already invested a significant amount in R&D and innovation either directly through Innovate UK and its predecessors or through underpinning research funded by the Research Councils. However, as we move to more challenging times and are faced with greater opportunities, it is essential that both government and the financial sectors as well as the food and drink industry increase their investment. This increased investment fits with the other aspects of the 5i approach – a skilled workforce who are able to translate ideas into innovation and are supported by improved infrastructure and increased investment as the vital ingredient to hold the mixture together.

Chemistry – The Enabling Science

Chemistry has been and continues to be the key enabling science that will underpin progress in the agri-food space. Both challenges and opportunities will continue to be identified and met with new and innovative science and collaborations. Meaningful engagement of the academic, policy, industry and regulatory stakeholders will be essential to the future health, wealth and prosperity of the sector.

EDITORS CONTACT INFORMATION

Professor A. Bryan Hanley
Specialist, Agri Food,
Knowledge Transfer Network, London, UK
Email: Bryan.hanley@ktn-uk.org

Dr. Wendy R. Russell
Senior Research Fellow,
University of Aberdeen Rowett Institute,
Aberdeen, Scotland, UK

INDEX

#

5-A-DAY, 191

A

acrylamide, 63, 252
additive manufacturing, x, xii, xiv, 3, 10, 194, 196, 201, 202, 232
adulteration, 54, 58, 79, 80, 89, 223
agricultural policy, 24, 25, 28
agronomy, xii, 18, 32, 49, 224
anaerobic digestion, 55, 203, 210, 217
antibiotics, 18, 20, 32, 57, 64
anti-browning, 131, 135
antimicrobial, 18, 32, 131, 132, 133, 134, 136, 138, 140, 171, 175, 189, 222, 235
appearance, 43, 64, 130, 144, 192, 239, 240, 244, 245
appetite, 41, 59
aquaculture, 17, 32, 99, 104, 226
attributes, 37, 61, 239, 244, 245

B

bioactives, ix, 151, 163
biotechnology, vi, 34, 105, 106, 111, 115, 124, 171, 184, 194, 210, 214, 217, 219, 221, 222, 223, 224, 225, 231, 232, 233, 234, 235
breakdown, 40, 41, 47, 242, 243, 244, 246
Brexit, 24
Browning, 134

C

cell walls, 38, 39, 46, 47, 50
chemical, 3, 9, 14, 21, 22, 29, 37, 39, 40, 45, 55, 61, 70, 74, 78, 79, 82, 84, 85, 88, 89, 90, 92, 95, 96, 97, 98, 101, 104, 105, 115, 116, 119, 121, 124, 130, 131, 134, 135, 136, 137, 138, 144, 145, 148, 150, 151, 155, 156, 157, 170, 171, 187, 203, 209, 210, 212, 214, 217, 218, 224, 238, 239, 241
chemical technologies, 29, 137, 138
chirality, 231, 232

climate change, viii, 14, 15, 19, 25, 28, 33, 54, 92, 96, 100, 102, 107, 111, 114, 116, 123, 154, 207, 238
Common Agricultural Policy (CAP), 24
consumer acceptance, 142, 197, 238
convenience foods, 10, 194, 197
co-products, 104, 154, 157, 159, 208
CRISPR, 17, 198, 230
CRISPR-Cas9, 17, 198, 230
crop protection, xiii, 20, 97, 99, 105, 197, 198, 199, 200, 227

D

dichlorodiphenyl trichloroethane (DDT), 2, 21, 22
dietary fibre, 142, 154, 155, 156, 157, 212
diet-related disease, viii, 195
digestible, 39, 41, 48
Docosahexadienoiic acid (DHA), 7

E

encapsulation, 52, 151, 152, 177
European Food Safety Authority (EFSA), 7, 27, 34, 35, 63, 86, 88, 119, 155, 174
evolution, 25, 32, 38, 50, 221, 245

F

fast food, 192
fat, 37, 39, 41, 58, 59, 120, 122, 139, 141, 143, 144, 145, 146, 147, 149, 152, 153, 158, 159, 173, 174, 175, 176, 178, 179, 199, 233, 234, 241
fat mimetics, 144, 146, 147
fat substitutes, 144, 145, 174, 175
fermentation, 3, 4, 5, 9, 17, 41, 46, 47, 116, 140, 141, 214, 215, 218, 221, 222, 223
fertiliser, 2, 15, 20, 24, 56, 94, 99
fetal programming, 8
flavours, x, 60, 61, 75, 76, 77, 82, 90, 136, 140, 141, 146, 151, 152, 173, 179, 196, 203, 222, 231, 232, 235, 239, 240, 241, 244, 245
Flavr Savr, 196
food industry, vii, ix, xiv, 11, 53, 54, 59, 61, 63, 64, 65, 66, 76, 77, 79, 80, 91, 102, 137, 146, 151, 152, 154, 163, 169, 171, 172, 177, 183, 201, 217, 221, 223, 230, 248, 249, 253, 254, 255
food processing, vi, xii, 3, 9, 10, 24, 26, 38, 39, 42, 44, 50, 94, 117, 124, 129, 130, 164, 166, 187, 190, 195, 201, 202, 203, 208, 212, 221, 231, 235, 250
food reformulation, 142, 152, 154, 172
food security, 13, 19, 28, 32, 33, 40, 43, 49, 67, 104, 113, 118, 122, 129, 164, 165, 194, 195, 205, 226, 250
food structure, 38
formulation, 98, 122, 139, 142, 178, 214, 241, 244
fortification, 118, 122, 147, 148, 149, 150, 151, 169, 176, 177

G

global economy, 10, 13
global warming, 6, 18, 54, 93, 116
golden rice, 197
green revolution, 20, 34
grey water, 55

H

Haber-Bosch, 2
health and nutrition, 154

I

insect protein, 63, 204, 226
insects, x, 2, 7, 16, 21, 23, 31, 34, 43, 63, 67, 94, 106, 120, 199, 200, 204, 226, 229, 233
iodine, 117, 131, 132, 148, 150, 177
iron, 45, 51, 117, 121, 138, 148, 149, 176, 230

K

kinetics of release, 241

L

liking, 2, 237, 238, 239
livestock production, 15, 16, 18, 31, 93, 97, 100, 109
long chain polyunsaturated fatty acids (LC-PUFAs), 7

M

mastication, 233, 241, 242, 243, 244, 246
meat, vii, x, 7, 16, 17, 18, 31, 38, 43, 44, 45, 50, 51, 62, 64, 80, 137, 138, 140, 141, 146, 147, 153, 171, 172, 173, 180, 187, 192, 194, 204, 222, 225, 227, 233, 237
microbiological load, 134
microbiome, xiii, 10
micronutrient deficiency, 117
myco-protein, 44
mycotoxins, 72, 83, 115, 124

N

natural climate solutions, 15
natural resources, 14, 24, 28, 31, 66, 104, 107, 207
NHANES, 193
novel processing technology, 201
nutrient, 18, 19, 28, 33, 41, 42, 48, 55, 59, 78, 99, 105, 115, 116, 117, 118, 137, 164, 224

O

obesity, viii, 29, 38, 39, 41, 42, 43, 45, 46, 48, 50, 53, 59, 125, 126, 144, 154, 193, 235, 238
over-nutrition, 113, 118

P

perception, xiii, 75, 121, 173, 196, 239, 243, 245, 246
pest damage, 199, 229
pesticide, 2, 21, 22, 23, 28, 34, 35, 64, 86, 94, 97, 200, 228
pesticides, 2, 15, 20, 21, 22, 23, 27, 34, 35, 72, 74, 75, 82, 89, 94, 114, 120, 227, 228
physical techniques, 138
physicochemical properties, 40, 43, 49, 173, 181, 182
phytochemical, 119, 120, 161
plant breeding, 124, 224
population growth, vii, 3, 5, 44, 54, 113, 117
preservation, 5, 8, 9, 17, 24, 39, 40, 62, 114, 130, 136, 137, 143, 144, 171, 191, 221, 222, 223, 235
primary agriculture, v, 13, 14
primary production, xiv, 3, 6, 8, 26, 49, 154, 204, 208, 224
processing, xiii, xiv, 1, 8, 11, 15, 38, 39, 40, 41, 42, 43, 45, 49, 51, 62, 63, 72, 83, 84, 85, 92, 95, 104, 105, 115, 116, 122, 129, 130, 134, 137, 138, 156, 157, 160, 161, 162, 163, 164, 165, 166, 168, 170, 171, 181, 183, 184, 185, 186, 187, 188, 195,

197, 201, 202, 204, 208, 210, 212, 215, 216, 217, 218, 221, 223, 231, 237, 243, 246, 248, 252
Producer Support Estimate (PSE), 26
protein, ix, 4, 7, 8, 10, 16, 28, 37, 40, 43, 44, 45, 58, 59, 62, 63, 80, 84, 85, 106, 118, 120, 121, 126, 141, 143, 144, 146, 152, 156, 157, 158, 159, 160, 178, 182, 183, 184, 185, 186, 194, 195, 199, 201, 204, 212, 214, 225, 226, 227, 230, 231, 233, 240, 242, 254

R

receptors, 2, 23, 195, 240, 241, 242, 245
reformulation of food, 122

S

saliva, 48, 243, 244
salt, 9, 10, 40, 59, 60, 120, 122, 133, 139, 140, 141, 143, 149, 150, 159, 171, 172, 173, 184, 242
satiety, 47, 59, 172
soil ecosystem, 116
starch, 39, 40, 46, 47, 48, 52, 95, 121, 132, 144, 146, 147, 150, 153, 155, 157, 178, 208, 218
sugar, 4, 5, 9, 40, 42, 51, 59, 60, 61, 67, 95, 101, 120, 122, 135, 139, 141, 142, 143, 144, 150, 152, 161, 208, 228, 233
sugar replacement, 60, 143
sustainability, v, 20, 24, 25, 30, 33, 91, 97, 98, 101, 103, 107, 120, 124, 125, 127, 138, 164, 190, 197, 202, 207, 226, 234, 244
sustainable food, 106, 112, 120, 121, 190
sweeteners, 60, 141, 142, 143, 231
synthetic meat, 233

T

taste, x, xiii, 2, 10, 43, 62, 63, 76, 140, 141, 144, 195, 196, 231, 232, 234, 239, 240, 241, 242, 243
texture(s), 10, 38, 41, 42, 43, 54, 60, 61, 62, 63, 75, 140, 141, 143, 144, 147, 173, 186, 239, 240, 242, 243, 244, 245, 246
thrifty baby syndrome, 8, 45

U

under-nutrition, 38, 114, 118

V

Vitamin A, 78, 149, 176

Z

Zoonotic diseases, 18

β

β-Carotene, 149

The Pineapple: Production, Utilization and Nutritional Properties

Editor: Lydia Hampton

Series: Food and Beverage Consumption and Health

Book Description: The opening review included in *The Pineapple: Production, Utilization and Nutritional Properties* discusses the fundamental and applied aspects related to production of aroma compounds in pineapple fruit.

Softcover ISBN: 978-1-53614-594-6
Retail Price: $82

Wine and Winemaking: The Value and Efficiency

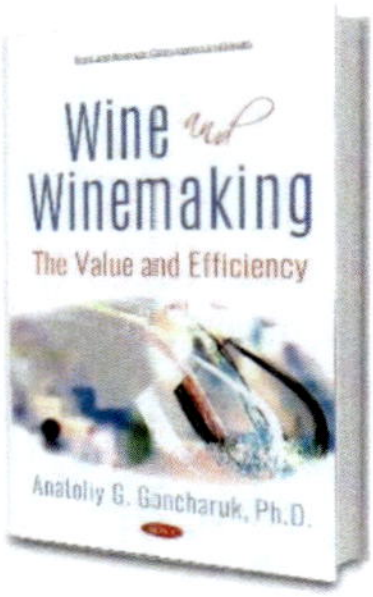

Author: Anatoliy G. Goncharuk, Ph.D. (Head of Department of Business Administration and Corporate Security, International Humanitarian University, Odessa, Ukraine)

Series: Food and Beverage Consumption and Health

Book Description: The author of this book wanted to know what the value of wine in today's society is, whether it has changed since ancient times, what kind of people consume wine today, and whether the desired effect of winemaking for its producers and consumers has been achieved.

Hardcover ISBN: 978-1-53613-385-1
Retail Price: $160

Fruit and Vegetable Consumption and Health: New Research

Editor: Noboru Motohashi (Meiji Pharmaceutical University, Kiyose-shi, Tokyo, Japan)

Series: Food and Beverage Consumption and Health

Book Description: It is becoming clear that incorporating vegetables and fruits into everyday meals is essential for human health maintenance from many experiments. This book mainly describes the effects, preventions and treatments of phytochemicals.

Hardcover ISBN: 978-1-53613-885-6
Retail Price: $195

Bread: Consumption, Cultural Significance and Health Effects

Editor: Hilda Lewis

Series: Food and Beverage Consumption and Health

Book Description: Bread is a pivotal food in many cultures around the world. Besides energy, it provides some nutritional benefits, being an important source of dietary fibres and vitamins.

Hardcover ISBN: 978-1-53612-274-9
Retail Price: $160